Advanced Topics in Science and Technology in China

Volume 4

Zhejiang University is one of the leading universities in China. In Advanced Topics in Science and Technology in China, Zhejiang University Press and Springer jointly publish monographs by Chinese scholars and professors, as well as invited authors and editors from abroad who are outstanding experts and scholars in their fields. This series will be of interest to researchers, lecturers, and graduate students alike.

Advanced Topics in Science and Technology in China aims to present the latest and most cutting-edge theories, techniques, and methodologies in various research areas in China. It covers all disciplines in the fields of natural science and technology, including but not limited to, computer science, materials science, the life sciences, engineering, environmental sciences, mathematics, and physics.

This book series is indexed by both the **Scopus** and **Compendex** databases.

If you are interested in publishing your book in the series, please contact Violetta Xu (Email: violetta_xuqian@zju.edu.cn) and Mengchu Huang (Email: mengchu.huang@cn.springer.com).

Honghao Lyu · Geng Yang · Huayong Yang

Human Motion Awareness and Robot Teleoperation

Perception, Communication and Control

ZHEJIANG UNIVERSITY PRESS
浙江大学出版社

Springer

Honghao Lyu
Zhejiang University
Hangzhou, China

Geng Yang
Zhejiang University
Hangzhou, China

Huayong Yang
Zhejiang University
Hangzhou, China

ISSN 1995-6819 ISSN 1995-6827 (electronic)
Advanced Topics in Science and Technology in China
ISBN 978-981-96-6544-0 ISBN 978-981-96-6545-7 (eBook)
https://doi.org/10.1007/978-981-96-6545-7

Jointly published with Zhejiang University Press
The print edition is not for sale in China mainland. Customers from China mainland please order the print book from: Zhejiang University Press.

This Springer imprint is published by the registered company Springer Nature Singapore Pte Ltd.
The registered company address is: 152 Beach Road, #21-01/04 Gateway East, Singapore 189721, Singapore

To my family, and all dreamers in robotics.

Honghao Lyu
Hangzhou, China

Preface

Teleoperation is a human-in-the-loop technology for controlling robots remotely. The human operators are responsible for high-level planning and cognitive decision-making, while the robot is responsible for manipulation. This allows humans to control the robot remotely without having to be physically present on-site. However, it is difficult for non-expert human operators like healthcare workers to master the control skills of a teleoperated robot in a short time using traditional teleoperation methods. With the development of motion capture and human-robot interaction technologies, human-motion-based teleoperation has become a research focus in the field. However, the mechanisms of human-robot motion mapping strategies are unclear, leading to difficulties in dual-arm motion control and full-body motion control of the robots with high degree-of-actuation. Additionally, the network latency and uncertainty of the communication within the control loop affect the control performance of teleoperation systems. To address these issues, the research of the book works on four aspects: single-arm motion mapping, dual-arm motion mapping, full-body motion mapping, and reliable communication for teleoperation. Moreover, two application cases are conducted to validate the usability of the human motion mapping strategies from the perspective of remote home care and remote medical assistance in Healthcare 4.0. The main research findings are summarized as follows:

(1) A human-motion-based mapping framework for single-arm teleoperation is proposed. The framework uses the operator's upper limb movements as input to control the end effector of the robotic arm. A path-constrained mapping method is proposed to improve the motion trajectory tracking performance of the robot during the teleoperation process. Based on this method, the position tracking deviation between the trajectory of the operator and robot is 1.05 mm at the sampling frequency of 7.5 Hz for simple motion trajectories, while the position tracking deviation is 5.10 mm for complex trajectories.
(2) An incremental motion mapping strategy is proposed to eliminate the cumbersome initialization and calibration process typically required in other motion mapping teleoperation systems. It allows flexible adjustments of the operator's

limb posture, improving the convenience and intuitiveness of the teleoperation. Furthermore, a hybrid mapping technique of hand gesture and upper-limb motion (GuLiM) is presented as a novel approach in dual-arm human-robot motion mapping. Comparative experiments showed that the GuLiM method surpassed the Directly Mapping Method (DMM) at the placement precision for position transfer and orientation transfer, with an improvement in accuracy of 46.77% and 69.27%, respectively.

(3) A hand gesture trajectory recognition method is proposed using background feature extraction and speed analysis. 10 types of gesture commands are designed for controlling robot locomotion. The recognition algorithm reaches 97.34% recognition accuracy in cross-validation. Further, a trunk motion mapping teleoperation strategy is introduced based on the upper body spine motion and lower limb motion of the operator, achieving intuitive control of the robot trunk with redundant degrees of freedom. Two application cases of elderly care and medical assistance in Healthcare 4.0 are conducted to verify the proposed human-motion-based teleoperation in the indoor environment and hospital isolation ward, respectively.

(4) A network hardware-in-the-loop simulation framework is developed to evaluate the impacts of the network performance on remote robot control. The interplay between communication and control is investigated under conditions such as 5G and Wi-Fi 6 local network connections, and a selection rule is given for choosing the wireless networks in local areas with different profiles. Moreover, a cloud-based teleoperation framework is introduced over ultra-long-distance wide-area network connections. A novel feedforward controller is designed to reduce tracking errors between the operator and the robot. This approach has successfully enabled intercontinental human-motion-based teleoperation over 7800 km, from Sweden to China.

Hangzhou, China
July 2024

Honghao Lyu
Geng Yang

Contents

List of Figures

List of Tables

Chapter 1
Introduction

Abstract This chapter begins by addressing the question "What is Robot Teleoperation". It introduces various application scenarios of robot teleoperation, such as remote homecare and remote medical assistance, citing relevant research findings. The chapter then discusses key technical points for teleoperation in healthcare, highlighting the limitations of traditional single-arm industrial robots, which are unsuitable for complex remote nursing tasks. In contrast, dual-arm collaborative robots offer greater flexibility and coordination, making them more suitable for nursing assistance that requires two-handed operations. The chapter also emphasizes the role of human motion capture technology in enhancing remote control intuitiveness and reducing training costs. Despite significant advancements, many technical challenges remain, particularly in terms of interaction intuitiveness and network performance. The chapter further reviews the evolution and categorization of teleoperation systems, covering the key concepts of human-motion-based teleoperation, human–robot motion mapping, and communication technologies. It discusses open challenges in teleoperation systems, such as accurately transferring human poses to robots, achieving smooth movement paths, and ensuring dual-arm and whole-body coordination. Finally, the chapter explores the prospects for next-generation teleoperated assistive robots, highlighting the need for improved strategies in limb pose mapping, dual-arm operation, and control-network co-design, particularly for unstructured caregiving tasks. The chapter concludes with an overview of the book's structure.

Keywords Teleoperation · Robot teleoperation applications · Healthcare robotics · Teleoperation background · Human-motion based control

1.1 What is Robot Teleoperation?

Robot teleoperation is a type of remote robot control technology with human operators in the loop. The human operator is responsible for high-level planning and cognitive decision-making, while the robot itself executes the corresponding control

H. Lyu et al., *Human Motion Awareness and Robot Teleoperation*, Advanced Topics in Science and Technology in China 4, https://doi.org/10.1007/978-981-96-6545-7_1

commands [1, 2]. Teleoperation enables robots to harness human high-level decision-making abilities to perform various tasks in complex and dynamic environments, thereby expanding the operational capabilities of robots [3, 4]. At the same time, it spares human users from having to be directly present in the robot's operational environment, allowing them to remotely control the robot from a local setting, thereby safeguarding human safety. For example, in extremely hazardous environments such as deep-sea exploration [5, 6], deep space exploration [7, 8], nuclear power plant maintenance [9, 10], and chemical waste disposal [11], teleoperation robots have significant application value [12, 13]. The development of intelligent perception technology [14], intelligent control technology [15], network transmission technology [16, 17], and the emergence of collaborative robots have further expanded the application scenarios of teleoperation technology. This has enabled teleoperation robots to have new applications in fields such as remote homecare and remote medical assistance [18–23].

1.1.1 Emerging Applications of Robot Teleoperation in Healthcare

Remote homecare: With the continuous increase in the global proportion of elderly populations, population aging has become a common challenge faced by countries worldwide [24, 25]. It is projected that by 2050, the proportion of people aged 60 and above in the world's total population will reach 21%, with the elderly population in the European Union and the United States reaching 28 and 22% respectively. As of 2015, China's population aged 60 and above has reached 220 million, accounting for 16% of China's total population. By 2050, this number is expected to reach 500 million, accounting for one-third of China's total population [26]. With the increasing elderly population, the societal burden of elderly care is growing, inevitably increasing the demand for caregivers. However, the current number of caregivers is limited and far from sufficient to meet the needs of an aging society [27, 28]. Teleoperation robots can remotely provide professional technical support from caregivers to complex, unstructured home environments with multiple user terminals [27, 29, 30], alleviating the shortage of homecare personnel. This represents a crucial solution to the challenge of elderly care [31, 32].

Remote medical assistance: Public health emergencies such as the COVID-19 pandemic have triggered major crises in global life, health security, and socio-economic order [33]. During previous major outbreaks, a significant number of front-line healthcare workers were infected [34, 35]: During the SARS outbreak in 2003, 1002 healthcare workers in mainland China were infected, accounting for 18.8% of all cases [36]; in the Middle East Respiratory Syndrome (MERS) outbreak, 106 healthcare workers were infected, accounting for 13.5% of all cases [37]. Teleoperation robots, as a form of remote human–robot interaction system, can act as a 'second body' for healthcare workers. They can assist or replace healthcare workers in entering isolation wards to perform diagnosis and care operations [36, 38–40]. This reduces the number of contacts between healthcare workers and patients, lowers the risk of nosocomial infections, effectively protects the lives and health of healthcare workers [41–44], and alleviates the issue of nosocomial infections among frontline healthcare workers.

1.1.2 Key Technical Points for Teleoperation in Healthcare

Traditional industrial robots are typically single-armed, with operational capabilities restricted to simple tasks, lacking intrinsic intelligence. They are ill-suited for performing complex remote nursing assistance operations effectively. Dual-arm collaborative robots exhibit higher coordination and flexibility, offering significant advantages in tasks requiring two-handed participation [45]. They can play a more substantial role in nursing assistance tasks characterized by high dynamics and strong environmental interactions, executing teleoperation tasks more flexibly and accurately conveying the intentions of caregivers [46]. On the other hand, human motion capture technology can intuitively interpret operators' motion intentions without additional training costs, enabling remote intuitive teleoperation with humans in the loop [47]. Despite years of research by industry researchers on teleoperation technology, there are still many technical challenges in terms of interaction intuitiveness, convenience, and other aspects. The impact of teleoperation loop network performance on control performance has not been fully understood. Improving existing teleoperation methods or proposing more sophisticated interactive methods has always been a key focus in the development of robot teleoperation systems [48]. Currently, for the implementation of robot teleoperation nursing assistance applications, the following urgent technical requirements are identified:

The intuitiveness of remote control is crucial in the interaction process of robot teleoperation. Convenience of operation and intuitive control are indispensable factors. Since the operators of nursing assistant robots are professional medical personnel, their time is limited. Traditional teleoperation input devices have cumbersome interfaces and high training costs, which cannot meet the convenience requirements of medical staff for operations [49]. Furthermore, medical personnel perform complex professional actions, and traditional interaction input methods struggle to encode the true operational intentions of medical staff, thus hindering effective coordination between robots and operators [50]. Therefore, it is imperative to conduct in-depth research on human–robot pose mapping methods based on operator limb motion data to enhance the intuitiveness of remote control of robots by medical personnel.

The convenience of dual-arm operation. Currently, nursing assistant robots in unstructured operating environments such as isolation wards, large hospitals, and home settings typically have a single-arm structure, which cannot meet the operational requirements of complex diagnostic and therapeutic assistance tasks [51]. Dual-arm manipulation allows for more flexible execution of complex tasks.

However, in practical operations, frequent switching between control targets and coordinating the left and right arms to complete tasks poses challenges. Traditional dual-arm teleoperation systems designed based on conventional input devices are costly, with complex interfaces that do not meet the intuitive and convenient control requirements for dual-arm teleoperation [52]. Therefore, it is essential to conduct in-depth research on robot dual-arm pose mapping teleoperation control methods to meet the complex teleoperation task requirements in nursing assistant scenarios.

The coordination of full-body movements. In recent years, to meet the task demands of complex, unstructured operational scenarios such as homecare and medical assistance, robot designs have evolved beyond arm actuators. They typically include mobile bases and redundant degrees of freedom mechanisms like humanoid torsos to enhance adaptability to complex operational tasks [53, 54]. Traditional teleoperation methods designed for single-arm robots are inadequate for coordinating such redundant degrees of freedom mechanisms [55, 56]. Therefore, there is an urgent need to research full-body coordinated teleoperation methods based on operator limb motion data. This will enable full-body coordinated teleoperation of humanoid robots with redundant degrees of freedom in high-dynamic, intense environmental interaction scenarios.

The reliability of network design. Communication networks is an indispensable component of teleoperation systems. Local wireless networks and wide-area wired networks enhance the deployment flexibility of robotic system, allowing operators to remotely control robots over different spatial distances [57]. However, compared to local wired networks, wireless and wide-area networks introduce additional transmission delays and communication unreliability, which significantly impact latency-sensitive teleoperation control systems [58–60]. Therefore, there is an urgent need for systematic research on the impact of network communication performance on robot teleoperation control performance [61]. It is crucial to further analyze the correlation between network performance and control performance in teleoperation systems, aiming to achieve coordinated design of network and control systems for pose mapping teleoperation systems.

Due to complex environments and diverse tasks, teleoperated nursing assistant robots still face many scientific challenges and technological bottlenecks in intuitive control, dual-arm mapping, collaborative operations, and network communication. Responding to these scientific challenges and the technological requirements in practical applications, this study focuses on key technologies for pose mapping teleoperation and intelligent interaction between humans and robots. The study explores single-arm and dual-arm pose mapping methods based on operator limb motion data input, as well as pose mapping teleoperation control methods for robot mobile bases and humanoid torsos with redundant degrees of freedom mechanisms. It investigates the impact mechanism of network performance on control performance in teleoperation control systems to enhance the intuitiveness, collaboration, and reliability of robot teleoperation technology. Furthermore, the study aims to apply these advancements to new application scenarios of teleoperation technology in remote homecare and remote medical assistance under the concept of Health 4.0. This research contributes to advancing the application of nursing assistant robots in China, accelerating the rapid development of the nursing assistant robot industry by providing crucial technological support.

1.2 Cutting-Edge Research in Robot Teleoperation

1.2.1 Overview and Classification

Teleoperation systems primarily refer to robotic systems where operators use master input devices to control slave robot for exploration, accomplishing tasks and so on. Communication between the master device and the slave robot is facilitated through network connections [2], and it is significant to realize high reliability and low latency in teleoperation systems. About remote human–robot interaction with next-generation network technologies, Professor Gerhard P. Fettweis from Dresden University of Technology has proposed the concept architecture of the Tactile Internet [62–64]. The IEEE 1918.1 Tactile Internet working group [65] has unified the definition of next-generation teleoperation systems into three parts: the master domain, network domain, and slave domain [66], as shown in Fig. 1.1. The master domain primarily includes the operator and the operating system interface. The slave domain consists of remote operating devices directly controlled by various command signals from the master domain, enabling interaction with various objects in the remote environment and providing feedback information back to the master domain. The network domain serves as a medium for bidirectional control communication between the master domain and the slave domain. Under the drive of technologies such as 5G or Ultra Reliable Low Latency Communications (URLLC), it dynamically couples operators to remote environments.

Traditional teleoperation applications are mostly oriented towards space exploration [67], deep-sea exploration [68], and nuclear power plant maintenance [69]. The earliest teleoperation systems can be traced back to 1954, when C. Goertz and others at the Argonne National Laboratory (ANL) developed two symmetrical master–slave mechanical arms for handling nuclear waste tasks, with the slave arm replicating the movements of the master arm through mechanical structure [70]. Additionally, in the 1980s, NASA's Jet Propulsion Laboratory (JPL) developed a dual-arm operating system for simulating teleoperation in deep space [71]. In 1993, the German Aerospace Center (DLR) conducted the first actual remote test in space with the ROTEX robot teleoperation system mounted on the Space Shuttle Columbia [72] (Fig. 1.2).

As technology advances, teleoperation techniques have begun to be applied in the fields of remote diagnosis, remote surgical assistance, and other aspects of healthcare [73, 74]. As shown in Fig. 1.3, in 2001, the teleoperated robotic system ZEUS developed by Computer Motion Inc. enabled the first demonstration of remote-assisted gallbladder removal surgery conducted from France [75]. Additionally, in 2012,

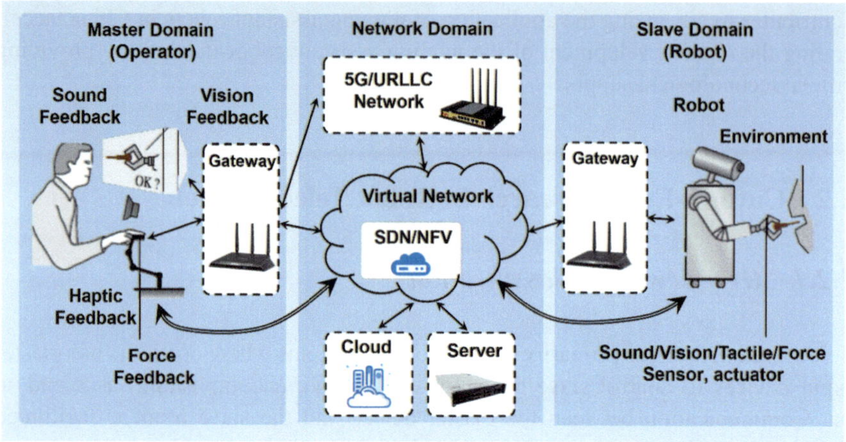

Fig. 1.1 The architecture of teleoperation systems driven by the tactile internet

(a) A teleoperation system for nuclear waste handling [70]

(b) A teleoperation system for deep space exploration [71]

(c) A teleoperation system carried on board the space shuttle [72]

Fig. 1.2 Tracing the applications of traditional teleoperated robotic systems. **a** A teleoperation system for nuclear waste handling [70]. **b** A teleoperation system for deep space exploration [71]. **c** A teleoperation system carried on board the space shuttle [72]

supported by teleoperation technology, the DaVinci surgical robot system developed by Intuitive Surgical Inc. successfully completed a minimally invasive heart surgery [76].

Tsinghua University [77, 78], Southeast University [79], Shanghai Jiao Tong University [80], and Harbin Institute of Technology [81] have all conducted research on key technologies in the field of teleoperated robotics. As shown in Fig. 1.4a, Harbin Institute of Technology has developed a satellite on-orbit autonomous maintenance teleoperation system. This system belongs to a typical bilateral teleoperation system, utilizing force feedback gloves and a spatial mouse as interaction devices at the near end. It has also established a three-dimensional virtual predictive environment for satellite maintenance teleoperation experiments [81]. The successful implementation

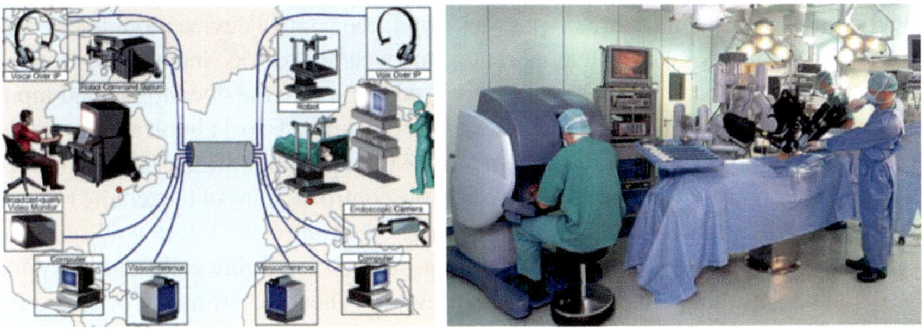

(a) The first transoceanic remote surgery demonstration [75]

(b) The Da Vinci Robot Surgical System [76]

Fig. 1.3 The application of teleoperated robotic systems in the healthcare field. **a** The first transoceanic remote surgery demonstration [75]. **b** The Da Vinci robot surgical system [76]

of China's Chang'e-5 lunar lander's lunar soil sampling mechanical arm in the third phase of the lunar exploration program, as well as the mechanical arm on the "Yutu" lunar rover in the second phase, as depicted in Fig. 1.4b, signifies the increasing strength of China's space technology and marks a milestone in its technological development [82].

Most of the teleoperation systems mentioned above are customized devices tailored for specific special applications or tasks, lacking generality. With the development of control and communication technologies, many universal teleoperation control input devices are emerging for robot teleoperation systems. According to the structural and functional characteristics of input devices, this book categorizes

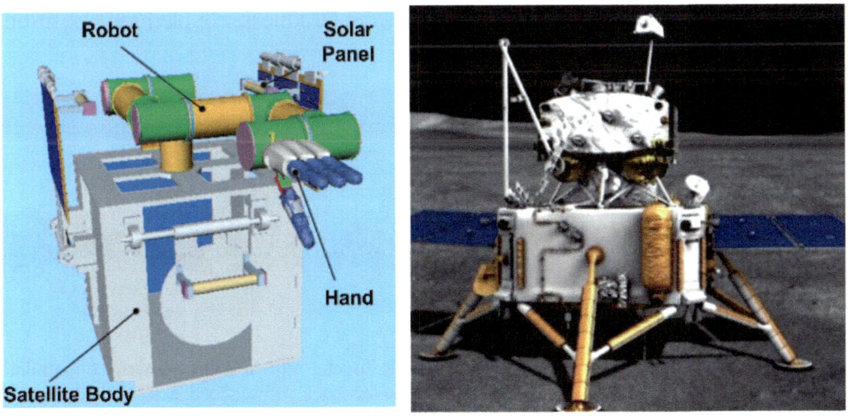

(a) The satellite in-orbit autonomous maintenance teleoperation system [81]

(b) The Chang'e-5 lander [82]

Fig. 1.4 Research progress of teleoperated robot systems in China. **a** The satellite in-orbit autonomous maintenance teleoperation system [81]. **b** The Chang'e-5 lander [82] teleoperation systems into three types: desktop force control device input teleoperation, exoskeleton device input teleoperation, and wearable device input teleoperation. Desktop force feedback devices are mainly used to capture the operator's hand input motions for teleoperating the end effector of robotic arms. Exoskeleton and wearable devices can capture the overall movements of the operator's limbs or even movements of other body parts, enabling more flexible control actions of the remote robotic system through human–robot motion mapping.

Desktop force control teleoperation input devices are lightweight and easy-to-install universal human–robot interaction devices. They are primarily operated by the operator gripping them with their hands to provide teleoperation inputs, also known as hand controllers [83]. The hand controller, while capturing the operator's end-effector motion information, also provides feedback forces from the slave side. Structurally, it can be categorized into series and parallel types. As shown in Fig. 1.5a, the team led by Song Aiguo from Southeast University developed a heterogeneous 6-degree-of-freedom force feedback hand controller for teleoperation experiments inside a spacecraft. This controller increases the workspace and feedback forces, thereby enhancing control precision [84, 85]. As shown in Fig. 1.5b and c, currently, one of the representative devices is the series of serial hand controllers from the American company 3D Systems, known as Geomagic Touch [86]. As well as the parallel force feedback hand controllers from the Swiss company Force Dimension, known as the Delta/Omega [87].

Exoskeleton teleoperation input devices are a kind of master input devices for remote operation tasks that follow the movements of the operator's limbs, mostly consisting of rigid serial link mechanisms [88, 89]. Based on the configuration differences with the controlled robot at the slave end, they can be classified into homogenous and heterogeneous exoskeletons. A seven-degree-of-freedom upper limb exoskeleton developed by the University of Washington features joints and linkages adapted to human body, but suffers from bulkiness, coupled joint positions

(a) The heterogeneous 6-degree-of-freedom force feedback hand controller [84]

(b) The Geomagic Touch series of serial force feedback haptic devices [87]

(c) The Omega series of parallel force feedback haptic devices by Force Dimension [87]

Fig. 1.5 Representative desktop force feedback teleoperation input devices. **a** The heterogeneous 6-degree-of-freedom force feedback hand controller [84]. **b** The geomagic touch series of serial force feedback haptic devices [87]. **c** The omega series of parallel force feedback haptic devices by force dimension [87]

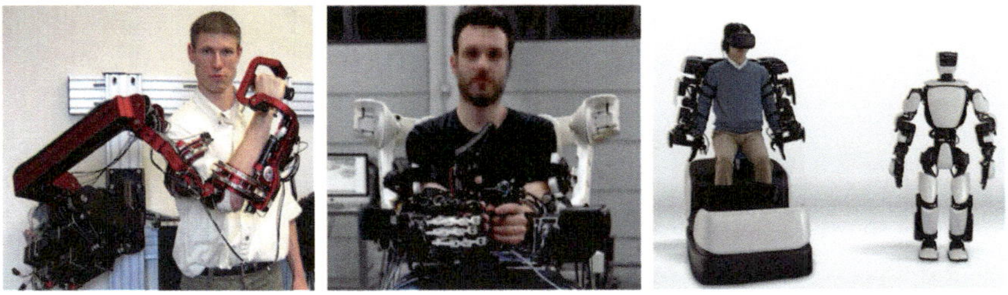

(a) 7-DoF exoskeleton [90]

(b) An exoskeleton device capable of capturing coupled movements of the upper limbs and wrists [91]

(c) Remote exoskeleton device for operating Toyota's T-HR3 robot [92]

Fig. 1.6 Representative exoskeleton teleoperation input devices. **a** 7-DoF exoskeleton [90]. **b** An exoskeleton device capable of capturing coupled movements of the upper limbs and wrists [91]. **c** Remote exoskeleton device for operating Toyota's T-HR3 robot [92]

and torques, and a complex drive system, leading to lower motion control precision, as depicted in Fig. 1.6a [90]. An exoskeleton developed by the University of Pisa in Italy for centaur robot teleoperation [91] flexibly captures upper limb and wrist movements of the operator, shown in Fig. 1.6b, yet it is bulky (weighing 2.8 kg), difficult to wear, and lacks naturalness in human–robot interaction. A more mature product currently is Toyota's T-HR3 robot's master control exoskeleton [92], as shown in Fig. 1.6c, enabling operators to wear the exoskeleton and control the movement of the humanoid robot.

Wearable teleoperation input devices refer to various devices worn on the human body to capture actions of the operator at the master end for controlling robots [93]. Current technological implementations mainly include electromyographic (EMG) signal acquisition [94, 95], inertial motion capture [96, 97], and wearable optical marker motion capture [98, 99]. As shown in Fig. 1.7a, a wearable EMG armband developed by Yang's team at Zhejiang University captures muscle signals for teleoperating a dexterous robotic hand [100]. In Fig. 1.7b, the University of Freiburg in Germany uses the wearable inertial motion capture device MVN to help the humanoid robot Nao mimic full-body actions of human [101]. In addition to these new interaction devices, wearable optical marker motion capture devices, as depicted in Fig. 1.7c, enable precise tracking of human movements. Researchers at École Polytechnique Fédérale de Lausanne in Switzerland used this technology to achieve intuitive teleoperation of a quadcopter drone [102].

Summarizing the technical characteristics of the aforementioned three types of teleoperation input devices, as shown in Table 1.1, it can be seen from the review of the teleoperation robot systems that traditional teleoperation robot systems mainly use desktop force control input devices at the master end, such as multi-degree-of-freedom hand controllers with customized mechanical structures. With the innovation in sensor technology, various new human–robot interaction devices have emerged in recent years, significantly enhancing the ability to intuitively capture the operator's

(a) A wearable electromyographic (EMG) armband [100]

(b) Wearable inertial motion capture devices [101]

(c) Wearable optical marker-based motion capture devices [102]

Fig. 1.7 Representative wearable teleoperation input devices. **a** A wearable electromyographic (EMG) armband [100]. **b** Wearable inertial motion capture devices [101]. **c** Wearable optical marker-based motion capture devices [102]

Table 1.1 Summary of technical characteristics of teleoperation input devices

Classification	Operator activity space	Device weight	Limb motion capture capability
Desktop force control teleoperation devices	Fixed	Fixed to desktop	No
Exoskeleton teleoperation devices	Mostly fixed	Heavy	Yes
Wearable teleoperation devices	Free to move	Lightweight	Yes

motion intentions [103]. Functionally, new interactive input devices represented by exoskeleton input devices and wearable motion capture devices can flexibly capture the limb movements and postures of the master end operator, achieving more flexible human–robot posture mapping for teleoperation interaction.

1.2.2 Human-Motion-Based Teleoperated Robot Systems

With the development of various new motion capture devices and human–robot interaction technologies, teleoperation based on the mapping of operator limb movements has enhanced the intuitiveness and convenience of controlling remote robots. This has become a research hotspot in the current field. This section reviews the frontier research work in the field of teleoperation based on limb movement mapping.

In 2020, a research team from Carnegie Mellon University (CMU) developed a vision-based teleoperation system called DexPilot [53]. This system uses four fixed RealSense depth cameras to capture the movements of the operator's hands and fingers, mapping and transmitting these movements to a 7-degree-of-freedom collaborative robotic arm (KUKA LBR iiwa7 R800) and a four-fingered dexterous hand (Allegro Hand) [104], achieving intuitive and dexterous grasping, as shown in Fig. 1.8a. By obtaining depth image (RGBD) information, the system uses deep neural networks to estimate the position and posture of the operator's hand. It then uses inverse kinematics to solve for the target joint angles of the robotic arm, redirecting the end-effector pose of the robotic arm. Furthermore, based on the open-source finger motion tracking model DART (Dense Articulated Real-Time Tracking) [105], it achieves dexterous movement mapping of finger postures, as shown in Fig. 1.8b.

Researchers at the Massachusetts Institute of Technology (MIT) developed a humanoid robot named HERMES that can imitate human movements and maintain balance [106]. To control HERMES, they designed a custom exoskeleton device capable of capturing the operator's limb movements and sensing the operator's standing balance [107], as shown in Fig. 1.9a. This device not only maps the operator's limb movements to the robot to control its joints but also senses the operator's balance state through the operator's waist movements and posture, adjusting the robot's balance accordingly. Additionally, through posture feedback from the robot, the exoskeleton can feed the humanoid robot's balance state back to the operator, achieving closed-loop control. The model diagram of this exoskeleton device is shown in Fig. 1.9b.

Researchers from the Technical University of Catalonia (UPC) have proposed a framework for human–robot pose mapping [108]. In this work, researchers use the Xsens wearable inertial motion capture devices [109] to capture the operator's full-body movements in real-time and construct a corresponding skeletal model. They also propose a variable admittance controller to use the acquired human poses for teleoperation control of the TIAGo robot, as shown in Fig. 1.10a. This system not only enables teleoperation of the robotic arm based on hand poses but also achieves

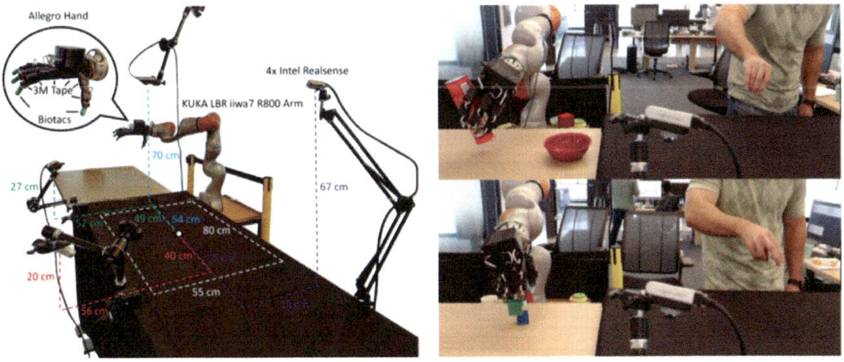

(a) DexPilot System [53]

(b) DexPilot teleoperated grasping demonstration [53]

Fig. 1.8 DexPilot teleoperation system developed by CMU. **a** DexPilot system [53]. **b** DexPilot teleoperated grasping demonstration [53]

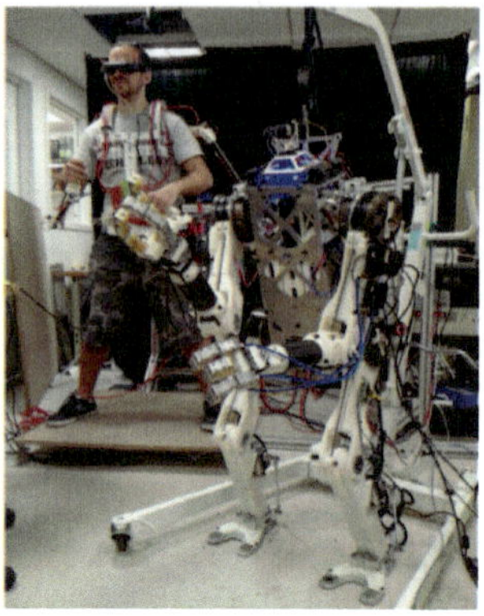

(a) HERMES Robot and its teleoperation device [106]

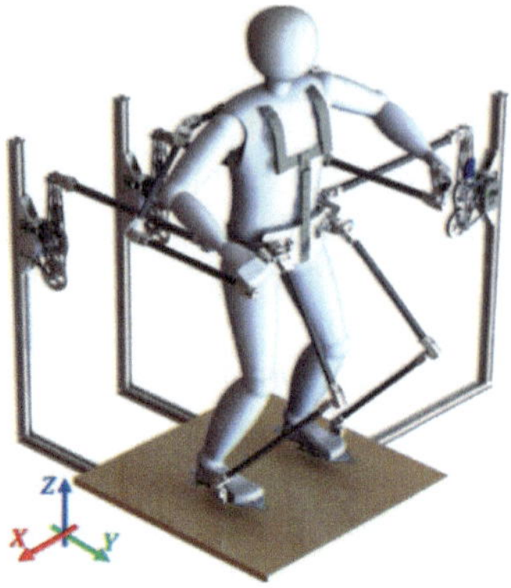

(b) Model diagram of the balance feedback exoskeleton device [107]

Fig. 1.9 The exoskeleton teleoperation system for HERMES robot from MIT. **a** HERMES Robot and its teleoperation device [106]. **b** Model diagram of the balance feedback exoskeleton device [107]

teleoperation of the chassis based on limb spatial displacement and control of the TIAGo robot's lift degree of freedom using a torso pose mapping strategy. The teleoperation process is illustrated in Fig. 1.10b.

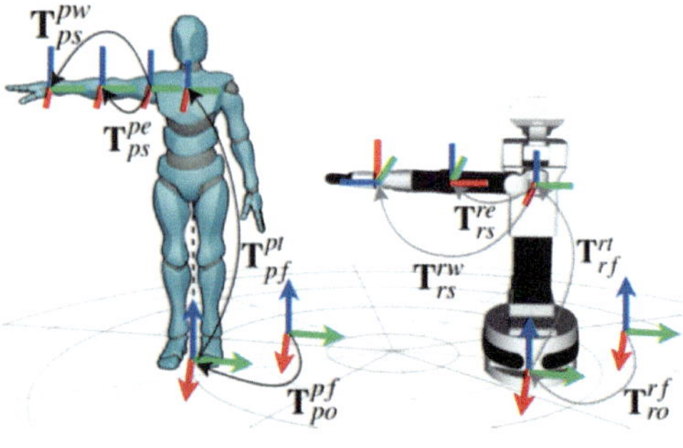

(a) Whole-body skeletal modeling and motion pose mapping strategy [109]

(b) Teleoperation demonstration [109]

Fig. 1.10 Teleoperation system based on inertial motion capture from UPC. **a** Whole-body skeletal modeling and motion pose mapping strategy [109]. **b** Teleoperation demonstration [109]

1.2.3 Human–Robot Motion Mapping Methods

The teleoperation of redundant degrees of freedom robots is more complex compared to traditional robotic arms. Examples include multi-degree-of-freedom robotic arms [110], dual-arm collaborative robots [45], and humanoid robots with redundant degrees of freedom such as chassis and torso movements. A straightforward approach to controlling such robots is capturing operator movements and mapping them to the robot's control system. However, the motion structure of these robots typically differs from that of the operator's limb structure, which make it necessary to do research into methods for mapping limb poses obtained from the operator to the robot's motion space. Existing methods for human–robot pose mapping primarily include three approaches [111, 112]: joint-to-joint mapping [113, 114], pose mapping [115, 116], and point-to-point mapping [117, 118]. This section provides a review of these three methods and summarizes the classification and technical characteristics of pose mapping methods in Table 1.2.

Joint-to-joint mapping refers to a mapping method where each sensor measuring the operator's motion is directly associated with the motion control of each robot joint. This mapping method is suitable for teleoperation scenarios where the slave-end robot and the operator have similar kinematic structures. If there is a clear correspondence between the joints of the human and the robot, the joint angles of the human can be directly mapped to the robot's joint control with minimal configuration scaling or direct mapping such as customized exoskeleton teleoperation systems designed for specific robot joint configurations. This mapping method offers significant advantages in high-load teleoperation tasks involving grasping, but it requires the slave-end robot to have a humanoid kinematic structure. Researchers at Tsinghua University, led by Fang Bin, utilized this method by capturing specific joint angles of the operator using an Inertial Measurement Unit (IMU) and directly mapping the collected joint motion information to the Baxter robotic arm, achieving joint-to-joint mapping teleoperation [119], as shown in Fig. 1.11

Table 1.2 Summary of classification and technical characteristics of human–robot pose mapping methods

Pose mapping methods	Applicable mapping configurations	Kinematic solution	Applicable operators
Joint-to-joint mapping	Master–slave isomorphic	Not required	Robotic arms
Pose mapping	Master–slave isomorphic/ heterogeneous	Required	Robotic arms
Point-to-point mapping	Master–slave isomorphic/ heterogeneous	Required	Robotic arms, dexterous hands

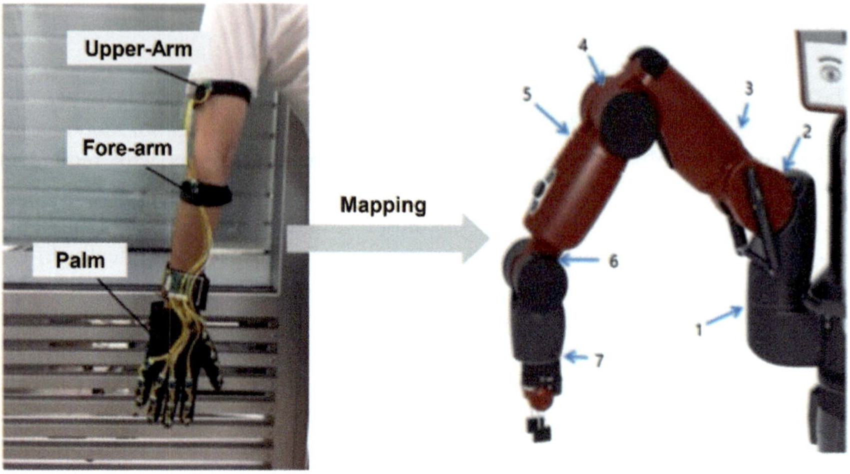

Fig. 1.11 Teleoperation of the Baxter robot based on joint-space mapping [119]

Pose mapping refers to a method that corresponds the operator's pose to the pose in the robot's coordinate space [120]. Pose mapping attempts to replicate human grasping functionality by mimicking the fingers' pose during operator grasping processes, thereby achieving teleoperation of human–robot functionality transfer rather than simply replicating the position of the operator's hands. However, the transformation in pose mapping is not perfectly precise, and pose mapping may result in unpredictable movements of the slave robot, restricting current research to simpler operational scenarios. Pose mapping typically involves theories related to machine learning and pattern recognition. Current research largely focuses on generating strategies for target poses and mapping rules, with more studies in human–robot "demonstration learning" and "imitation learning" [121, 122]. Yet, research and applications of real-time teleoperation based on pose mapping remain relatively limited.

Point-to-point mapping refers to a method where the position of specified points on the operator's arm or hand is replicated to control designated points on the robot to reach corresponding positions in its coordinate space. This method is commonly used for pose mapping in teleoperating dexterous hand robots [117], and it is also frequently employed when precise positioning of the robot's end effector is required [123]. This method involves processing human limb pose data and performing forward kinematic calculations to determine the position of the operator's hand. Subsequently, inverse kinematic calculations are used to derive the spatial state of the robot's joint. Point-to-point mapping typically involves underactuated or redundant degree-of-freedom forward and inverse kinematic solutions. Due to differences in the kinematic structures of operators and robots, issues such as no solution in inverse kinematics or significant discrepancies between the inverse kinematic solution and the operator's configuration may arise. These challenges make general point-to-point teleoperation less intuitive, requiring additional training and adaptation time for operators to correctly operate the system. Moreover, in environments where the remote robot system faces non-structural, constrained, or obstacle-rich conditions, operators are prone to errors under high workload pressures [124].

1.2.4 Communication Technologies for Robot Teleoperation

The introduction of local wireless networks has increased the flexibility of robotic system, simplifying network hardware design and installation processes compared to wired network connections [59]. This has also reduced the maintenance requirements for communication hardware to some extent, particularly beneficial for mobile robots with large spatial movement requirements [59, 125], enabling remote teleoperation of robots [60, 126]. However, due to the open and dynamic propagation environment of electromagnetic waves, the shared transmission medium, and uncertain performance of actual wireless network, instability existsin remote robot control [127]. In addition, wide-area wired network connections cover extensive ranges, often spanning multiple cities, countries, or even globally, enabling operators to remotely control robots over varying spatial distances. As remote operation distances increase, the expansive topology and complex network management of wide-area networks pose challenges such as significant latency and control instability for teleoperation systems [128, 129].

Zhang et al. from Nanjing Medical University have developed a remote robotic arm-assisted ultrasound diagnostic system based on 5G mobile communication technology. They applied this system during medical missions in remote pastoral areas of Xinjiang, where access to medical care is limited. The feasibility and qualitative performance of the system were validated primarily from the perspectives of ultrasound imaging quality and network latency. However, quantitative analysis of performance metrics related to teleoperation control of the robotic arm and real-time network performance was not conducted [130]. As shown in Fig. 1.12, the Joint Japanese-French Robotics Laboratory (JRL) used a wide-area network connection to preliminarily achieve remote teleoperation control of a 6-degree-of-freedom single-arm HRP-2 robot from Germany to Japan using a self-built master–slave operating lever. Yet, a comprehensive approach integrating remote network performance and control system design for teleoperation systems has not been fully developed [131].

To ensure the reliable operation of robotic systems, it is essential during the design and deployment phases of industrial robot platforms to simulate and test the system's communication performance [132, 133]. Introducing delays or other communication characteristics into controllers to simulate network performance is a common solution [134]. However, simulating the statistical performance of real-world network conditions, especially under long-term operation of wireless networks, can be challenging [135]. Li et al. simulated the wireless sensor network performance for unmanned aerial vehicle (UAV) swarm control in Industrial Internet of Things (IIoT) applications using the V-REP simulation software [136]. Although this simulation method

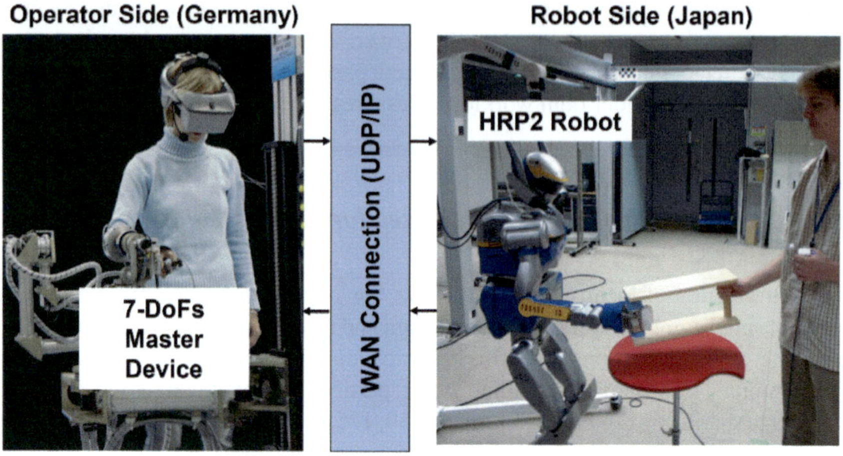

Fig. 1.12 Teleoperation of HRP-2 robot in Japan from Germany [131]

allows for long-term stability testing, it does not assess the general capability to handle different network conditions and communication uncertainties. Parameter-based network simulation methods are often used to validate the coordination of Automated Guided Vehicles (AGVs) [137, 138], but they generally do not systematically evaluate the uncertainties introduced by network communication into control systems.

Hardware-in-the-Loop (HiL) simulation refers to a simulation technique where a portion of a physical system is simulated using computer models, while physical devices are directly interfaced to participate in the simulation [139–142]. Scholars worldwide have conducted extensive research on HiL simulation for robot systems [141–144]. Zhou et al. proposed a HiL method for simulating acoustic communication, discussing its potential application in underwater robot systems, but without testing it on robot devices [139, 141]. Marco et al. introduced a HiL system for pre-validation and optimization of tasks for space station robotic arms on the ground, assuming real-time computation and zero communication delay [142–145]. Lamping et al. designed a multi-agent UAV system based on the Robot Operation System (ROS) [144] and used HiL simulation to test control algorithms in a multi-vehicle system [145, 146]. However, existing HiL simulation studies in robot applications often focus on controller design and do not address communication system design within robotic systems [147, 148]. Markus et al. from ABB and Nokia utilized a communication simulation method based on traffic models to study the support capabilities of 5G and LTE for industrial robot field applications [58, 60]. However, their simulation results are limited to predefined task scenarios and do not evaluate the impact of network performance and uncertainties on control performance. A common assumption in HiL simulations for robot control systems is ideal communication conditions, particularly for robotic systems under wireless network control

Table 1.3 Summary of research on communication systems for robot control

Research category		Reference number	Long-term testing capability	Generalization across different application scenarios	Evaluation of communication uncertainty
Parameter model-driven simulation methods	Underwater robots	[138]	Yes	Low	None (ideal communication assumption)
	Autonomous navigation vehicles	[139, 140]	No	Low	None
Other hardware-in-the-loop (HiL) simulation testing methods	HLSP/ UWAC	[142]	Yes (only for communication)	Low	Yes
	HiL-OC	[145]	No	Low	None (ideal communication assumption)
	5G-robotics	[60]	Complex configuration and implementation	Low	None
	N-HiL [148]	[150]	Yes	Moderate	Yes (only applicable to Wi-Fi 6)

[146, 149]. This book provides a comparative summary of current research on control and communication co-design in robot control systems, as shown in Table 1.3.

1.3 Open Challenges and the Prospects

1.3.1 Challenges for Human-Motion-Based Teleoperation

Research on Limb Pose Mapping Teleoperation Systems faces challenges: Due to differences in motion and configuration between robots and human bodies, accurately transferring operator motion poses to the robot is a major challenge in pose mapping teleoperation. Traditional teleoperation processes aim to improve tracking accuracy by increasing control frequency, but this often results in jerky and non-smooth movement paths, reducing the overall system's operational fluidity and efficiency. Decreasing the output frequency during pose mapping seems an intuitive solution, yet ensuring similarity and tracking accuracy between human's and robot's end trajectories under low frequency control remains a fundamental issue in pose mapping processes.

Research on dual-arm Pose Mapping Teleoperation faces challenges: For complex tasks requiring the involvement of two arms, the master operator must control the robotic arms to move to target positions and manipulate end effectors to perform grasping actions. During practical operations, frequent switching between controlling objects and coordinating the left and right robotic arms is necessary to complete tasks efficiently. Traditional teleoperation solutions struggle to coordinate actions among different components, often requiring additional input devices for linked control, which burdens the master operator. Designing a convenient and efficient dual-arm pose mapping method is thus a critical challenge that urgently needs resolution for the widespread adoption of teleoperation technologies.

Challenges in Whole-Body Pose Mapping Teleoperation Research: Robots performing homecare or medical assistance tasks typically have mobile bases and humanoid torsos, in addition to robotic arm actuators, to control robot movement and adjust arm workspace. Traditional master control methods using joysticks and button operations are limited in their ability to meet the agile control demands of robots with high degrees of freedom in torso structures. The key technical challenge in pose mapping teleoperation applications lies in how to utilize the operator's limb motion poses to control actions of the robot's mechanisms other than the robotic arm actuators.

Challenges in Control and Communication Co-design for Teleoperation Systems: Pose mapping teleoperation tasks are often performed in complex and uncertain unstructured environments, demanding high requirements for communication latency and control reliability. However, the current understanding of how network performance affects teleoperation control systems remains unclear. Existing network communication architectures designed for teleoperation control systems are simplistic and fail to meet the complex task demands in latency-sensitive teleoperation scenarios. Therefore, the key challenge in current pose mapping teleoperation control processes lies in the control and communication co-design for robot control, specifically tailored for unstructured and complex caregiving assistance tasks.

1.3.2 Prospects of Human-Motion-Based Teleoperation

In the process of next-generation teleoperated assistive robot control, perception forms the foundation, communication stands as crucial, and control serves as the core. However, current research highlights numerous scientific challenges and technological bottlenecks in collaborative robot teleoperation applications for homecare and medical assistance scenarios: (1) The strategy for teleoperation control based on operator limb motion data and pose mapping remains unclear, resulting in awkward human–robot interaction, slow operator learning curves, and high training costs in teleoperating assistive robots for caregiving. (2) There is a lack of dual-arm pose mapping teleoperation strategies applied to unstructured caregiving tasks, limiting the dexterity required for complex operations by current caregiving assistive robots. (3) Methods for teleoperating redundant-degree-of-freedom caregiving assistive robots, including their mobile bases and humanoid torso structures, have not been adequately developed, restricting their remote operation and motion capabilities. (4) The mechanism by which network performance affects control performance in robot teleoperation control systems remains unexplored, resulting in the absence of control and network co-design strategies tailored for robot pose mapping teleoperation control.

Building upon these existing challenges and research gaps, this book integrates robot control technology with operator limb motion capture technology. It focuses on the application needs of teleoperation in unstructured task environments, leveraging the flexibility and intuitiveness inherent in human motion poses. The research aims to explore the impact mechanism of network performance on control performance, establish stable and reliable control network co-design strategies, and meet the convenient interaction requirements between operators and caregiving assistive robots. Ultimately, the goal is to transform teleoperated caregiving assistive robots into a remote "second body" for healthcare workers.

1.4 Organization of This Book

This book focuses on research into human–robot pose mapping teleoperation based on operator limb motion, covering four main aspects: single-arm pose mapping teleoperation, dual-arm pose mapping teleoperation control, full-body coordinated teleoperation, and control-network co-design reliability. The research is conducted with a specific focus on applications in Healthcare 4.0. By analyzing operator limb motion data, the study explores strategies for human–robot pose mapping teleoperation, methods for mapping and coordinating dual-arm actions, and strategies for full-body coordinated teleoperation based on operator torso motion data. Furthermore, the book investigates control and network co-design strategies for pose mapping teleoperation systems to achieve complex and dexterous operations of caregiving assistive robots in unstructured task environments. Based on the findings from dual-arm pose mapping and full-body coordinated teleoperation control, the research aims to validate the application of human–robot pose mapping technology in practical home and medical environments. Ultimately, the goal is to achieve technical milestones from "single-arm pose mapping" to "dual-arm pose mapping", further to "full-body coordinated teleoperation", and "reliable teleoperation with communication", supporting functional testing and application research in typical scenarios of remote home caregiving and telemedicine assistance in Healthcare 4.0.

1.4.1 Research Content

1. Research on Single-arm Human–Robot Pose Mapping Teleoperation Based on Limb Motion Data. The study involves inertia sensor data fusion techniques and human body posture calculation methods, constructing a kinematic model of the human skeleton based on joint chains, and reconstructing the human motion

posture model. It explores bone motion update transformation algorithms based on forward kinematics solutions, proposes a strategy for human–robot motion mapping based on the operator's hand posture, providing pose information input for motion control of the slave-end robot. The research establishes a robot single-arm kinematic model, explores real-time inverse kinematics resolution strategies for redundant degrees of freedom robotic arms, and achieves single-arm pose mapping teleoperation based on operator limb motion.

2. Research on dual-arm human–robot motion mapping teleoperation technology based on limb motion data. This involves studying the coordination control strategy between the dual arms of collaborative robots and dual end effectors, establishing a joint control method for robotic arms and end effectors for complex operation tasks, and developing real-time pose calculation and control enabling strategies for dual-arm redundant degree of freedom collaborative robots. It also explores the mechanism of joint-to-joint dual-arm pose mapping in the human–robot interaction process, designs an incremental pose mapping strategy based on master–slave real-time coordinate transformation, and achieves dual-arm pose mapping teleoperation based on the coupling of hand gestures and upper limb movements of the operator.
3. Design of a full-body coordinated pose mapping teleoperation system based on limb motion data. This involves researching methods for recognizing the dynamic trajectories of the operator's hand gestures and extracting features based on the operator's lower limb and torso movements. It proposes human–robot pose mapping strategies for redundant degree of freedom robots, such as mobile chassis and humanoid torso mechanisms, and explores full-body coordinated pose mapping methods. This research aims to provide technical support for full-body coordinated teleoperation of caregiving assistant robots in complex operation scenarios and to conduct application studies of pose mapping teleoperation in real home environments and medical assistance task scenarios.
4. Research on collaborative design methods for control and networks in human–robot pose mapping teleoperation. This research focuses on the reliability requirements of communication networks in teleoperation control systems, systematically exploring the impact mechanisms of network transmission delays and reliability on the control performance of robotic pose mapping teleoperation. It involves performance testing under local wireless networks and wide-area wired networks, analyzing the correlation between real-time network performance and control performance. The study investigates collaborative design and selection methods for networks based on different control performance requirements, providing technical support for reliable teleoperation control under complex network conditions and achieving reliable pose mapping teleoperation under various network connection conditions.

The technical route of this book revolves around various key scientific issues in human–robot single-arm pose mapping teleoperation, dual-arm pose mapping teleoperation, full-body coordinated teleoperation, and network and control collaborative design for remote care assistant robots. The research is conducted based on "deconstruction of scientific problems–design of key technical solutions–experimental verification and application", summarized as follows:

1. Key Technical Research 1—Single-arm Human–robot Pose Mapping Teleoperation Technology: By studying human–robot skeletal model reconstruction technology based on inertial motion capture, designing end effector trajectory tracking strategies for human–robot teleoperation, and developing real-time kinematic inversion methods for robotic arms, this research aims to achieve single-arm pose mapping teleoperation based on operator motion data.
2. Key Technical Research 2—Dual-arm Human–robot Pose Mapping Teleoperation Technology: This research focuses on dual-arm incremental pose mapping strategies, designing human–robot dual-arm action mapping teleoperation interfaces based on upper limb poses, and developing hybrid intuitive control strategies for dual-arm teleoperation based on operator gestures. The goal is to achieve dual-arm teleoperation based on upper limb motion pose mapping.
3. Key Technical Research 3—Full-body Coordinated Pose Mapping Teleoperation Technology: By studying mapping control strategies between the operator's torso and lower limb movements and humanoid robot redundant degree-of-freedom motion mechanisms, as well as distributed control operation logic, this research aims to achieve full-body coordinated pose mapping teleoperation based on operator limb movements.
4. Key Technical Research 4—Teleoperation Network and Control Collaborative Design Technology: This involves studying the impact mechanisms of real-time network communication performance on teleoperation control system performance, building teleoperation control architectures for local wireless networks and wide-area wired network connections, and achieving control and network collaborative design for pose mapping teleoperation systems.
5. Engineering Application Research—Remote Care Assistance: Based on the completion of the aforementioned key technical research, this research focuses on application requirements in the new scenario of Healthcare 4.0. It conducts application research and functional verification for two typical scenarios: remote homecare and remote medical assistance.

1.4.2 Structure of this Book

The organization of this book is illustrated in Fig. 1.13. Targeting the research goal of intuitive and convenient interaction between remote care assistant robots and operators, this study sequentially conducts research in four areas: single-arm human–robot pose mapping teleoperation, dual-arm pose mapping teleoperation, full-body coordinated teleoperation, and reliable communication teleoperation. Focusing on four key technologies: end trajectory tracking mapping, incremental dual-arm pose mapping, chassis-torso mapping control, and teleoperation control network collaborative design, the study identifies four critical scientific issues: (1) Intuitive Mapping

Teleoperation Strategy for End Pose of Human–Machine Interaction; (2) Mechanism of Dual-Arm Pose Mapping Teleoperation; (3) Full-Body Coordinated Pose Mapping Strategy; (4) Collaborative Design Method for Teleoperation Control Networks. These studies aim to improve the intuitiveness and convenience of remote interaction between teleoperation robots and operators. This book lays the theoretical and technical foundation for the practical application and promotion of the next-generation remote homecare robots and remote medical assistance robots, geared towards the new scenario of Healthcare 4.0 [149].

This book focuses on the study of human–robot pose mapping teleoperation technology based on body motion data. The research aims to provide technical support for achieving high-dynamic and strong-environment interaction tasks in care assistance through human–robot pose mapping teleoperation. The main contributions of this book are summarized as follows:

(1) Pose Mapping Methods for Robotic Arms Based on Body Motion Data: Addressing the unclear human–robot pose mapping relationship and the difficulty in coordinating end-effector control during the teleoperation of robots based on body motion. This research explores the end-effector trajectory

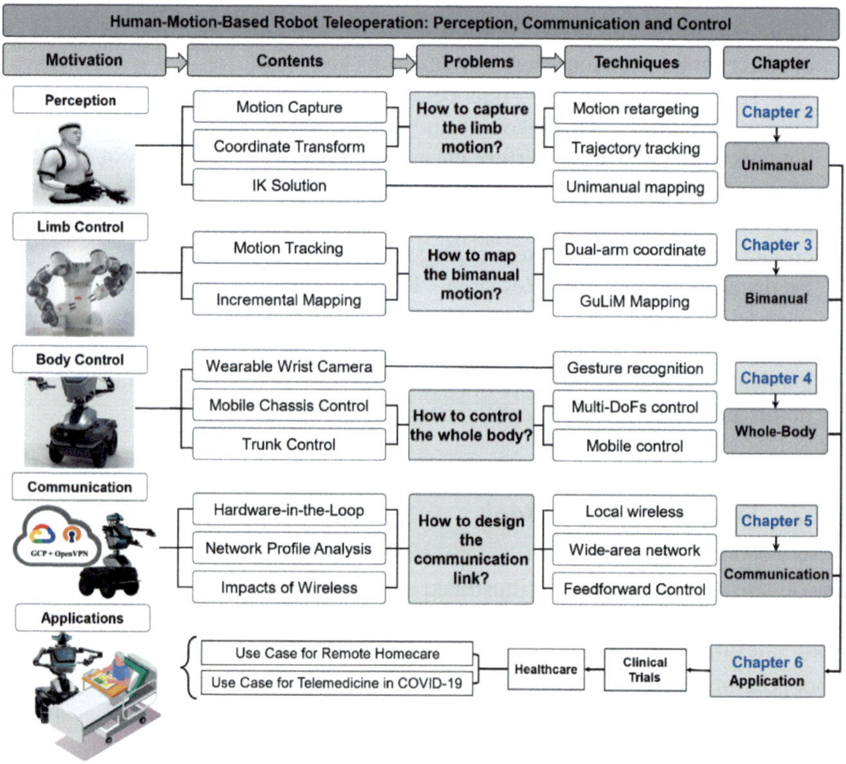

Fig. 1.13 The organizational structure of this book

tracking strategy in the human–robot motion mapping process, incremental pose mapping strategy based on real-time coordinate transformation, and the coordination control method of robotic arms and end-effectors during dual-arm teleoperation control. One of the main contributions of this dissertation is the realization of intuitive pose mapping teleoperation for robotic arms.

(2) Full-Body Coordinated Teleoperation Control Methods Based on Body Motion Data: Redundant degrees of freedom humanoid robots adjust the workspace of robotic arms through body motion mechanisms such as the chassis and torso, making full-body human–robot motion coordination challenging. This research investigates human body motion recognition and torso motion analysis strategies, exploring pose mapping strategies for the chassis and humanoid torso mechanisms based on operator body motion. Realizing full-body coordinated teleoperation is the second main contribution of this dissertation.

(3) Network-Control Coordination Design Methods for Pose Mapping Teleoperation Systems: Addressing complex network communication conditions in teleoperation control systems, this research tests long-term statistical performance indicators in delay-sensitive operation scenarios under new wireless network connection conditions, investigates the impact mechanism of network performance on robot control performance, and develops a set of network-control coordination strategies for different network characteristics to meet various control performance requirements. Achieving human–robot pose mapping teleoperation under different network connection conditions is the third main contribution of this dissertation.

References

1. Lai Y, Ye X, Ding H (2021) Research progress of major research plan on tri-co robots. Chin J Mech Eng 57(23):1–20
2. Siciliano B, Khatib O (2016) Springer handbook of robotics. Springer International Publishing, Cham, Switzerland
3. Zhang H, Zhou K, Shi K et al (2023) SmartSpring: a low-cost wearable haptic VR display with controllable passive feedback. IEEE Trans Visual Comput Graph 29(11):4460–4471
4. Shahbazi M, Atashzar SF, Tavakoli M et al (2018) Position-force domain passivity of the human arm in telerobotic systems. IEEE-ASME Trans Mechatron 23(2):552–562
5. Yang X, Yan J, Hua C et al (2021) Trajectory tracking control of autonomous underwater vehicle with unknown parameters and external disturbances. IEEE Trans Syst Man Cybern: Syst 51(2):1054–1063
6. Yu X, Qin HD, Zhu ZB (2023) Underwater localization of AUVs in motion using two-way travel time measurements with unknown sound velocity. IEEE Trans Veh Technol 72(9):11358–11373
7. Tang Q, Xia Q, Xu C et al (2022) Human-machine interaction system design for space-oriented manipulator teleoperation. J Command Control 8(03):278–285
8. Weihua LIJG, Liang DING, Haibo GAO (2023) State of art and prospects of ground teleoperation technology for lunar rover. Acta Aeronaut Astronaut Sin 44(1):26333–026333
9. De Geeter J, Decréton M, Colon E (1999) The challenges of telerobotics in a nuclear environment. Robot Auton Syst 28(1):5–17

10. Stedman H, Kocer BB, Zalk NV et al (2023) Evaluating immersive teleoperation interfaces: coordinating robot radiation monitoring tasks in nuclear facilities. In: Proceedings of the 2023 IEEE international conference on robotics and automation (ICRA), pp 11972–11978
11. Fischer GKJ, Bergau M, Gómez-Rosal DA et al (2024) Evaluation of a smart mobile robotic system for industrial plant inspection and supervision. IEEE Sens J 24(12):19684–19697
12. Wong C, Yang E, Yan X et al (2017) Adaptive and intelligent navigation of autonomous planetary rovers - a survey. In: Proceedings of the NASA/ESA conference on adaptive hardware and systems, pp 237–244
13. Jakuba MV, German CR, Bowen AD et al (2018) Teleoperation and robotics under ice: implications for planetary exploration. In: Proceedings of the IEEE aerospace conference, pp 1–14
14. Ha N, Xu K, Ren G et al (2020) Machine learning-enabled smart sensor systems. Adv Intell Syst 2(9):2000063
15. Hua Y, Zhang H, Li Y et al (2022) Vision assisted control of lower extremity exoskeleton for obstacle avoidance with dynamic constraint based piecewise nonlinear MPC. IEEE Robot Autom Lett 7(4):12267–12274
16. Yang G, Pang Z, Deen MJ et al (2020) Homecare robotic systems for healthcare 4.0: visions and enabling technologies. IEEE J Biomed Health Inform 24(9):2535–2549.
17. Hu F, Deng Y, Zhou H et al (2021) A vision of an XR-aided teleoperation system toward 5G/B5G. IEEE Commun Mag 59(1):34–40
18. Ding H, Yang X, Zheng N et al (2018) Tri-Co robot: a Chinese robotic research initiative for enhanced robot interaction capabilities. Natl Sci Rev 5(6):799–801
19. He Z, Wu J, Zhang J et al (2024) CDM-MPC: an integrated dynamic planning and control framework for bipedal robots jumping. IEEE Robot Autom Lett 9(7):6672–6679
20. Galsworthy M, McKee M (2013) Europe's 'horizon 2020' science funding programme: how is it shaping up? J Health Serv Res Policy 18(3):182–185
21. Zhang D, Xiao B, Huang B et al (2019) A self-adaptive motion scaling framework for surgical robot remote control. IEEE Robot Autom Lett 4(2):359–366
22. Tamantini C, Di Luzio FS, Cordella F et al (2021) A robotic healthcare assistant for the COVID-19 emergency: a proposed solution for logistics and disinfection in a hospital environment. IEEE Robot Autom Mag 28(1):71–81
23. Kennedy M (2023) The role of collaborative robotics in assistive and rehabilitation applications. Sci Robot 8(83):eadk6743
24. United Nations (1985) Department of international economic and social affairs., united nations. department for economic and social information and policy analysis. United Nations. Department of Economic and Social Affairs. Population Division. World population prospects. 1982–1990: population studies. New York, United Nations, Department of International, Economic and Social Affairs
25. Cheng X, Yang Y, Schwebel DC et al (2020) Population ageing and mortality during 1990–2017: a global decomposition analysis. PLoS Med 17(6):e1003138
26. Department of Economic and Social Affairs PD (2019) World population prospects. United Nations, New York
27. Bardaro G, Antonini A, Motta E (2022) Robots for elderly care in the home: a landscape analysis and co-design toolkit. Int J Soc Robot 14(3):657–681
28. Pouyan A, Panchea AM, François F (2022) A Review on the use of mobile service robots in elderly care. Robotics 11(6)
29. Liang X, Yan Y, Wang W et al (2024) Adaptive human–robot interaction torque estimation with high accuracy and strong tracking ability for a lower limb rehabilitation robot. IEEE/ASME Trans Mechatron 1–12
30. Hua Y, Zhu Y, Li C et al (2022) Assistance control of human-exoskeleton integrated system for balance recovery augmentation in sagittal plane. iEEE Trans Industr Electron 69(1):528–538
31. Melkas H, Hennala L, Pekkarinen S et al (2020) Impacts of robot implementation on care personnel and clients in elderly-care institutions. Int J Med Inform 134:104041

32. Li Z, Moran P, Dong Q et al (2017) Development of a tele-nursing mobile manipulator for remote care-giving in quarantine areas. In: Proceedings of the IEEE international conference on robotics and automation, Singapore, pp 3581–3586.
33. Sprung CL, Joynt GM, Christian MD et al (2020) Adult ICU triage during the coronavirus disease 2019 pandemic: who will live and who will die? Recommendations to improve survival. Crit Care Med 48(8):1196–1202
34. Bandyopadhyay S, Baticulon RE, Kadhum M et al (2020) Infection and mortality of healthcare workers worldwide from COVID-19: a systematic review. BMJ Glob Health 5(12):e003097
35. Chang D, Xu H, Rebaza A et al (2020) Protecting health-care workers from subclinical coronavirus infection. Lancet Respir Med 8(3):e13
36. Yang Y, Peng F, Wang R et al (2020) The deadly coronaviruses: the 2003 SARS pandemic and the 2020 novel coronavirus epidemic in China. J Autoimmun 109:102434
37. Chowell G, Abdirizak F, Lee S et al (2015) Transmission characteristics of MERS and SARS in the healthcare setting: a comparative study. BMC Med 13(1):210
38. Yang GZ, Nelson BJ, Murphy RR et al (2020) Combating COVID-19-the role of robotics in managing public health and infectious diseases. Sci Robot 5(40):eabb5589
39. Matsumoto S, Moharana S, Devanagondi N et al (2022) Iris: a low-cost telemedicine robot to support healthcare safety and equity during a pandemic. In: Proceedings of the pervasive computing technologies for healthcare, Cham, pp 113–133
40. Liu T, Liao Q h, Gan L et al (2021) The role of the Hercules autonomous vehicle during the COVID-19 pandemic: use cases for an autonomous logistic vehicle for contactless goods transportation. IEEE Robot Autom Mag 28(1):48–58
41. Editorial (2020) COVID-19: protecting health-care workers. Lancet, 395(10228):922
42. McCall B (2020) COVID-19 and artificial intelligence: protecting health-care workers and curbing the spread. Lancet Digital Health 2(4):166–167
43. Alenezi H, Cam ME, Edirisinghe M (2021) A novel reusable anti-COVID-19 transparent face respirator with optimized airflow. Bio-Des Manuf 4(1):1–9
44. Jovanovic K, Schwier A, Matheson E et al (2021) Digital innovation hubs in health-care robotics fighting COVID-19: novel support for patients and health-care workers across Europe. IEEE Robot Autom Mag 28(1):40–47
45. Abbas M, Narayan J, Dwivedy SK (2023) A systematic review on cooperative dual-arm manipulators: modeling, planning, control, and vision strategies. Int J Intell Robot Appl 7(4):683–707
46. Tavakoli M, Carriere J, Torabi A (2020) Robotics, smart wearable technologies, and autonomous intelligent systems for healthcare during the COVID-19 pandemic: an analysis of the state of the art and future vision. Adv Intell Syst 2(7):2000071
47. Fang B, Sun F, Liu H et al (2020) Wearable technology for robotic manipulation and learning. Springer Singapore, Singapore
48. Luo J, He W, Yang C (2020) Combined perception, control, and learning for teleoperation: key technologies, applications, and challenges. Cogn Comput Syst 2(2):33–43
49. Meeker C, Rasmussen T, Ciocarlie M (2018) Intuitive hand teleoperation by novice operators using a continuous teleoperation subspace. In: Proceedings of the IEEE international conference on robotics and automation, Brisbane, Australia, pp 5821–5827
50. Pang G, Yang G, Pang Z (2021) Review of robot skin: a potential enabler for safe collaboration, immersive teleoperation, and affective interaction of future collaborative robots. IEEE Trans Med Robot Bionics 3(3):681–700
51. Malik A A, Masood T, Kousar R (2020) Repurposing factories with robotics in the face of COVID-19. Sci Robot 5(43)
52. Makris S, Tsarouchi P, Surdilovic D et al (2014) Intuitive dual arm robot programming for assembly operations. CIRP Ann Manuf Technol 63(1):13–16
53. Handa A, Wyk KV, Yang W et al (2020) DexPilot: vision-based teleoperation of dexterous robotic hand-arm system. In: Proceedings of the IEEE international conference on robotics and automation, pp 9164–9170

54. Liu Y, Li Z, Su H et al (2021) Whole-body control of an autonomous mobile manipulator using series elastic actuators. IEEE/ASME Trans Mechatron 26(2):657–667
55. Munoz LM, Casals A (2009) Improving the human-robot interface through adaptive multispace transformation. IEEE Trans Rob 25(5):1208–1213
56. Pan S, Lyu H, Duan H et al (2019) A sensor glove for the interaction with a nursing-care assistive robot. In: Proceedings of the IEEE international conference on industrial cyber physical systems, pp 405–410
57. Park P, Ergen SC, Fischione C et al (2018) Wireless network design for control systems: a survey. IEEE Commun Surv Tutor 20(2):978–1013
58. Aleksy M, Dai F, Enayati N et al (2019) Utilizing 5G in industrial robotic applications. In: Proceedings of the international conference on future internet of things and cloud, pp 278–284
59. Baumann D, Mager F, Wetzker U et al (2021) Wireless control for smart manufacturing: recent approaches and open challenges. Proc IEEE 109(4):441–467
60. Maddahi Y, Liao S, Fung W et al (2015) Selection of network parameters in wireless control of bilateral teleoperated manipulators. IEEE Trans Industr Inf 11(6):1445–1456
61. Zhao G, Imran MA, Pang Z et al (2019) Toward real-time control in future wireless networks: communication-control co-design. IEEE Commun Mag 57(2):138–144
62. Fettweis GP (2014) The tactile internet: applications and challenges. IEEE Veh Technol Mag 9(1):64–70
63. Fanibhare V, Sarkar NI, Al-Anbuky A (2021) A survey of the tactile internet: design issues and challenges, applications, and future directions. Electronics 10(17):2171
64. Simsek M, Aijaz A, Dohler M et al (2016) 5G-enabled tactile internet. IEEE J Sel Areas Commun 34(3):460–473
65. Holland O, Steinbach E, Prasad RV et al (2019) The IEEE 1918.1 "tactile internet" standards working group and its standards. Proc IEEE 107(2):256–279
66. Gokhale V, Kroep K, Rao VS et al (2020) TIXT: an extensible testbed for tactile internet communication. IEEE Internet of Things Mag 3(1):32–37
67. McHenry N, Spencer J, Zhong P et al (2021) Predictive XR telepresence for robotic operations in space. In: Proceedings of the 2021 IEEE aerospace conference, vol 50100, pp 1–10
68. Phung A, Billings G, Daniele AF et al, Enhancing scientific exploration of the deep sea through shared autonomy in remote manipulation. Sci Robot 8(81):eadi5227
69. Liu M, Zhang Y, Zhu C et al (2016) Influence of time delay on team performance in space robotic teleoperation. Eng Psychol Cogn Ergon 9736(1):189–197
70. Goertz RC (1954) Mechanical master-slave manipulator. Nucleonics 12(11):45–46
71. Skaar SB, Ruoff CF (1994) Teleoperation and robotics in space. American Institute of Aeronautics and Astronautics, Washington, U.S.
72. Hirzinger G, Brunner B, Dietrich J et al (1993) Sensor-based space robotics - rotex and its telerobotic features. IEEE Trans Robot Autom 9(5):649–663
73. Anooshahpour F, Yadmellat P, Polushin IG et al (2019) A motion transmission model for a class of tendon-based mechanisms with application to position tracking of the da Vinci instrument. IEEE-ASME Trans Mechatron 24(2):538–548
74. Ballantyne GH (2002) Robotic surgery, telerobotic surgery, telepresence, and telementoring - review of early clinical results. Surg Endosc Other Interv Tech 16(10):1389–1402
75. Marescaux J, Leroy J, Rubino F et al (2002) Transcontinental robot-assisted remote telesurgery: feasibility and potential applications. Ann Surg 235(4):487–492
76. Gupta R, Tanwar S, Tyagi S et al (2019) Tactile-internet-based telesurgery system for healthcare 4.0: an architecture, research challenges, and future directions. IEEE Netw 33(6):22–29
77. Li S, Jiang J, Ruppel P et al (2020) A mobile robot hand-arm teleoperation system by vision and IMU. In: Proceedings of the IEEE/RSJ international conference on intelligent robots and systems, Las Vegas, USA, pp 10900–10906
78. Fang B, Wei X, Sun F et al (2019) Skill learning for human-robot interaction using wearable device. Tsinghua Sci Technol 24(6):654–662

79. Shen S, Song A (2019) Bilateral motion prediction and tracking control for nonlinear teleoperation system with time-varying delays. In: Proceedings of the international conference on control, automation and systems, Jeju, South Korea, pp 798–803
80. Rezazadeh S, Bai W, Sun M et al (2019) Robotic spinal surgery system with force feedback for teleoperated drilling. J Eng 2019(14):500–505
81. Jiang Z, Liu H, Wang J et al (2009) Virtual reality-based teleoperation with robustness against modeling errors. Chin J Aeronaut 22(3):325–333
82. Ma RQ, Jiang SQ, Liu B et al (2018) Design and verification of a lunar sampling manipulator system. J Astronaut 39:1315–1322
83. Adilkhanov A, Rubagotti M, Kappassov Z (2022) Haptic devices: wearability-based taxonomy and literature review. IEEE Access 10:91923–91947
84. Qin H, Song A, Liu Y et al (2015) Design and calibration of a new 6 DOF haptic device [J/OL] 2015 15(12):31293–31313. https://doi.org/10.3390/s151229857
85. Zhu C, Song A (2017) Position calibration of a 3-DOF hand-controller with hybrid structure. Rev Sci Instrum 88(9):095002
86. Hojati NS, Motaharifar M, Taghirad HD et al (2019) Skill assessment using kinematic signatures: geomagic touch haptic device. In: Proceedings of the international conference on robotics and mechatronics, pp 186–191.
87. Sanfilippo F, Weustink PBT, Pettersen KY (2015) A coupling library for the force dimension haptic devices and the 20-sim modelling and simulation environment. In: Proceedings of the annual conference of the IEEE industrial electronics society, Yokohama, Japan, pp 168–173
88. Mallwitz M, Will N, Teiwes J et al (2015) The capio active upper body exoskeleton and its application for teleoperation. In: Proceedings of the symposium on advanced space technologies in robotics and automation, pp 1–8
89. Planthaber S, Mallwitz M, Kirchner EA (2018) Immersive robot control in virtual reality to command robots in space missions. J Softw Eng Appl 11(7):341–347
90. Perry J C, Rosen J (2006) Design of a 7 degree-of-freedom upper-limb powered exoskeleton. In: Proceedings of the IEEE/RAS-EMBS international conference on biomedical robotics and biomechatronics, pp 805–810
91. Buongiorno D, Sotgiu E, Leonardis D et al (2018) WRES: a novel 3 DoF WRist ExoSkeleton with tendon-driven differential transmission for neuro-rehabilitation and teleoperation. IEEE Robot Autom Lett 3(3):2152–2159
92. Murray L (2021) Meet the Tokyo robots. Eng Technol 16(6):28–31
93. Ding X, Clifton D, Ji N et al (2021) Wearable sensing and telehealth technology with potential applications in the coronavirus pandemic. IEEE Rev Biomed Eng 14:48–70
94. Tortora S, Moro M, Menegatti E (2019) Dual-Myo real-time control of a humanoid arm for teleoperation. In: Proceedings of the ACM/IEEE international conference on human-robot interaction, pp 624–625
95. Khiabani H, Ahmadi M (2021) A classical machine learning approach for EMG-based lower limb intention detection for human-robot interaction systems. In: Proceedings of the IEEE international conference on autonomous systems, pp 1–5
96. Zhou H, Lyu H, Yi K et al (2019) An IoT-enabled telerobotic-assisted healthcare system based on inertial motion capture. In: Proceedings of the IEEE international conference on industrial informatics, Helsinki, Finland, pp 1373–1376
97. Lu C, Dai Z, Jing L (2023) Measurement of hand joint angle using inertial-based motion capture system. IEEE Trans Instrum Meas 72:9503211
98. Siaw T, Han Y, Wong K (2023) A low-cost marker-based optical motion capture system to validate inertial measurement units. IEEE Sens Lett 7(2):1–4
99. Chatzitofis A, Albanis G, Zioulis N et al (2022) A low-cost & realtime motion capture system. In: Proceedings of the IEEE/CVF conference on computer vision and pattern recognition, New Orleans, USA, pp 21421–21426
100. Yang G, Deng J, Pang G et al (2018) An IoT-enabled stroke rehabilitation system based on smart wearable armband and machine learning. IEEE J Transl Eng Health Med 6(1):2100510

101. Koenemann J, Bennewitz M (2012) Whole-body imitation of human motions with a nao humanoid. In: Proceedings of the ACM/IEEE international conference on human-robot interaction, Boston, USA, p 425
102. Macchini M, Schiano F, Floreano D (2020) Personalized telerobotics by fast machine learning of body-machine interfaces. IEEE Robot Autom Lett 5(1):179–186
103. Young SN, Peschel JM (2020) Review of human-machine interfaces for small unmanned systems with robotic manipulators. IEEE Trans Hum-Mach Syst 50(2):131–143
104. Guedri N, Gharbi R (2021) Finger movements tracking of the human hand using the smart camera to control the allegro hand robot. In: Proceedings of the IEEE international conference on signal, control and communication, Tunis, Tunisia, pp 320–324
105. Schmidt T, Newcombe RA, Fox D (2014) DART: dense articulated real-time tracking. In: Proceedings of the robotics: science and systems, Berkeley, USA, pp 1–9
106. Wang A, Ramos J, Mayo J et al (2015) The HERMES humanoid system: a platform for full-body teleoperation with balance feedback. In: Proceedings of the IEEE-RAS international conference on humanoid robots, Seoul, South Korea, pp 730–737
107. Ramos J, Wang A, Ubellacker W et al (2015) A Balance feedback interface for whole-body teleoperation of a humanoid robot and implementation in the HERMES system. In: Proceedings of the IEEE-RAS international conference on humanoid robots, Seoul, South Korea, pp 844–850
108. Arduengo M, Arduengo A, Colomé A et al (2019) A robot teleoperation framework for human motion transfer. arXiv:1909.06278
109. Drapeaux A, Carlson K (2020) A comparison of inertial motion capture systems: DorsaVi and Xsens. Int J Kinesiol Sports Sci 8(3):24–27
110. Lunardini F, Casellato C, d'Avella A et al (2016) Robustness and reliability of synergy-based myocontrol of a multiple degree of freedom robotic arm. IEEE Trans Neural Syst Rehabil Eng 24(9):940–950
111. Zheng Y, Tang Z, Dai JS (2018) A mapping method of grasping posture applying to the metamorphic multi-fingered hand. In: Proceedings of the international conference on reconfigurable mechanisms and robots, Delft, Netherlands, pp 1–7
112. Liarokapis M, Bechlioulis CP, Artemiadis PK et al (2017) Deriving humanlike arm hand system poses. J Mech Robot 9(1):011012
113. Kyriakopoulos KJ, Riper JV, Zink A et al (1997) Kinematic analysis and position/force control of the anthrobot dextrous hand. IEEE Trans Syst Man Cybern Part B (Cybern) 27(1):95–104
114. Liarokapis MV, Artemiadis PK, Kyriakopoulos KJ (2013) Mapping human to robot motion with functional anthropomorphism for teleoperation and telemanipulation with robot arm hand systems. In: Proceedings of the IEEE/RSJ international conference on intelligent robots and systems, Tokyo, Japan, pp 2075–2075
115. Fang B, Ma X, Wang J et al (2020) Vision-based posture-consistent teleoperation of robotic arm using multi-stage deep neural network. Robot Auton Syst 131:103592
116. Li S, Hendrich N, Liang H et al (2022) A dexterous hand-arm teleoperation system based on hand pose estimation and active vision. IEEE Trans Cybern 1–12
117. Peer A, Einenkel S, Buss M (2008) Multi-fingered telemanipulation - mapping of a human hand to a three finger gripper. In: Proceedings of the IEEE international symposium on robot and human interactive communication, Munich, Germany, pp 465–470
118. Gao Q, Li J, Zhu Y et al (2023) Hand gesture teleoperation for dexterous manipulators in space station by using monocular hand motion capture. Acta Astronaut 204:630–639
119. Fang B, Sun F, Liu H et al (2017) Robotic teleoperation systems using a wearable multimodal fusion device. Int J Adv Rob Syst 14(4):1–11
120. Li G, Caponetto F, Del Bianco E et al (2020) Incomplete orientation mapping for teleoperation with one DoF master-slave asymmetry. IEEE Robot Autom Lett 5(4):5167–5174
121. Ekvall S, Kragic D (2004) Interactive grasp learning based on human demonstration. In: Proceedings of the IEEE international conference on robotics and automation, New Orleans, USA, pp 3519–3524

122. Zhang T, McCarthy Z, Jow O et al (2018) Deep imitation learning for complex manipulation tasks from virtual reality teleoperation. In: Proceedings of the IEEE international conference on robotics and automation, Brisbane, Australia, pp 5628–5635
123. Pierce RM, Kuchenbecker KJ (2012) A data-driven method for determining natural human-robot motion mappings in teleoperation. In: Proceedings of the IEEE RAS & EMBS international conference on biomedical robotics and biomechatronics, Rome, Italy, pp 169–176
124. Muhayyuddin, Moll M, Kavraki L et al (2018) Randomized physics-based motion planning for grasping in cluttered and uncertain environments. IEEE Robot Autom Lett 3(2):712–719
125. Gielis J, Prorok A (2021) Improving 802.11p for delivery of safety-critical navigation information in robot-to-robot communication networks. IEEE Commun Mag 59(1):16–21
126. Fall CL, Quevillon F, Blouin M et al (2018) A multimodal adaptive wireless control interface for people with upper-body disabilities. IEEE Trans Biomed Circuits Syst 12(3):564–575
127. Lee S, Duong TQ, Woods R (2019) Impact of wireless backhaul unreliability and imperfect channel estimation on opportunistic NOMA. IEEE Trans Veh Technol 68(11):10822–10833
128. Manfredi S, Natalizio E, Pascariello C et al (2020) Stability and convergence of a message-loss-tolerant rendezvous algorithm for wireless networked robot systems. IEEE Trans Control Netw Syst 7(3):1103–1114
129. Baranitha R, Mohajerpoor R, Rakkiyappan R (2021) Bilateral teleoperation of single-master multislave systems with semi-Markovian jump stochastic interval time-varying delayed communication channels. IEEE Trans Cybern 51(1):247–257
130. Qijie Z, Jie Y, Maimitili A et al (2023) Application and exploration of 5G remote robotic ultrasound examination in network remote consultation and free consultation in border and pastoral areas. J Clin Ultrasound Med 25(02):85–89
131. Hirche S, Buss M (2012) Human-oriented control for haptic teleoperation. Proc IEEE 100(3):623–647
132. Kadian A, Truong J, Gokaslan A et al (2020) Sim2Real predictivity: does evaluation in simulation predict real-world performance? IEEE Robot Autom Lett 5(4):6670–6677
133. Cale J, Johnson B, Anese E et al (2018) Mitigating communication delays in remotely connected hardware-in-the-loop experiments. IEEE Trans Ind Electron 65(12):9739–9748
134. Liu Y, Lin T, Lin M (2022) Indirect/direct learning coverage control for wireless sensor and mobile robot networks. IEEE Trans Control Syst Technol 30(1):202–217
135. Herath JD, Seetharam A (2019) A Markovian model for analyzing opportunistic request routing in wireless cache networks. IEEE Trans Veh Technol 68(1):812–821
136. Li H, Savkin AV (2018) Wireless sensor network based navigation of micro flying robots in the industrial internet of things. IEEE Trans Ind Inf 14(8):3524–3533
137. Mannucci A, Pallottino L, Pecora F (2019) Provably safe multi-robot coordination with unreliable communication. IEEE Robot Autom Lett 4(4):3232–3239
138. Digani V, Hsieh MA, Sabattini L et al (2019) Coordination of multiple AGVs: a quadratic optimization method. Auton Robot 43(3):539–555
139. Zhou L, Liu L, Li M et al (2010) HLSP/UWAC: a hardware-in-loop simulation platform for underwater acoustic communication. In: Proceedings of the international conference on wireless communications networking and mobile computing, Chengdu, China, pp 1–4
140. Lyu H, Pang Z, Bhimavarapu K et al (2023) Impacts of wireless on robot control: the network hardware-in-the-loop simulation framework and real-life comparisons. IEEE Trans Industr Inf 19(9):9255–9265
141. Chen Y, Braun DJ (2019) Hardware-in-the-loop iterative optimal feedback control without model-based future prediction. IEEE Trans Rob 35(6):1419–1434
142. Stefano MD, Mishra H, Giordano AM et al (2021) A relative dynamics formulation for hardware in the loop simulation of on-orbit robotic missions. IEEE Robot Autom Lett 6(2):3569–3576
143. Kumar J, Amutha E (2014) Control and tracking of robotic manipulator using PID controller and hardware in loop simulation. In: Proceedings of the international conference on communication and network technologies, Sivakasi, India, pp 1–3

144. Quigley M, Conley K, Gerkey B et al (2009) ROS: an open-source robot operating system. In: Proceedings of the ICRA workshop on open source software, Kobe, Japan, p 5
145. Lamping AP, Ouwerkerk JN, Cohen K (2018) Multi-UAV control and supervision with ROS. In: Proceedings of the aviation technology, integration, and operations conference, Atlanta, p 4245
146. Hayat S, Yanmaz E, Muzaffar R (2016) Survey on unmanned aerial vehicle networks for civil applications: a communications viewpoint. IEEE Commun Surv & Tutor 18(4):2624–2661
147. Qi C, Gao F, Zhao X et al (2018) Distortion compensation for a robotic hardware-in-the-loop contact simulator. IEEE Trans Control Syst Technol 26(4):1170–1179
148. Lyu H, Pang Z, Xiao M et al (2022) Hardware-in-the-loop simulation for evaluating communication impacts on the wireless-network-controlled robots. In: Proceedings of the annual conference of the IEEE industrial electronics society, Brussels, Belgium, pp 1–6
149. Pang Z, Yang G, Khedri R et al (2018) Introduction to the special section: convergence of automation technology, biomedical engineering, and health informatics toward the healthcare 4.0. IEEE Rev Biomed Eng 11(1):249–259
150. Zahavi A, Haeri SN, Liyanage DC et al (2020) A dual-arm robot for collaborative vision-based object classification. In: Proceedings of the Biennial Baltic electronics conference, Tallinn, Estonia, pp 1–5

Chapter 2
Unimanual Human-Motion Based Robot Teleoperation

Abstract This chapter addresses the challenge of accurately mapping and controlling a robot based on the operator's body movements using a teleoperation system. The key issue lies in translating human body posture and intent into robot movements, considering the differences between human bodies and robots. Traditional methods improve tracking accuracy by increasing control frequency, but they can lead to uneven robot speeds and jerky motion. This study uses ABB's dual-arm collaborative robot, YuMi, and Noitom Perception Neuron (PN) motion capture device to track the operator's upper limb movements. The design involves a single-arm human pose mapping control interface and a distributed communication system between the robot and PN device. The primary challenge addressed is ensuring smooth and accurate end trajectory tracking despite low-frequency pose data output. To solve this, the chapter proposes a local path resampling algorithm. This method improves the tracking accuracy and trajectory similarity between the operator and robot, ensuring continuous trajectory mapping and tracking. Comparative experiments with varying resampling frequencies demonstrate that the best trajectory tracking occurs at a specific resampling rate, with minimal tracking errors in both simple (1.05 mm) and complex (5.10 mm) trajectories.

Keywords Human-motion based teleoperation · Single-arm motion mapping · Tracking accuracy · Robot operation system · Wearable motion capture

2.1 Introduction

It is an intuitive and convenient teleoperation method to map and control the robot by capturing the operator's body movements, in which the operator's body movements capture is an indispensable part of the pose mapping teleoperation system. In the actual teleoperation process, due to the differences between robots and the human bodies, how to accurately transmit the operator's motion posture and operation intention to the robot has become a major challenge. To ensure the similarity and tracking accuracy between the robot's end trajectory and the operator's end

H. Lyu et al., *Human Motion Awareness and Robot Teleoperation*, Advanced Topics in Science and Technology in China 4, https://doi.org/10.1007/978-981-96-6545-7_2

trajectory is a basic issue in the pose mapping process. Based on the robot's built-in kinematics solver, the tracking accuracy of traditional teleoperation system can be improved by increasing the control frequency, but this could lead to uneven speeds of the robot's end in the motion process, and the tracking path will be stuck and not smooth. This chapter takes ABB's dual-arm collaborative robot YuMi as the research platform [1], based on Neuteng's Perception Neuron (PN) motion capture device to capture the operator's upper limb movements [2], the control interface design of single-arm human pose mapping and the distributed communication architecture design between the robot and PN motion capture device are realized [3]. To solve the problem of uneven motion trajectory in the process of teleoperation is an intuitive solution, however, how to ensure the tracking accuracy of the end trajectory when low-frequency pose data mapping is output is a key challenge in teleoperation. In this chapter, a pose mapping method for end trajectory tracking based on local path resample constraints of operator end trajectory is proposed, which improves the tracking accuracy of master–slave end trajectory when low-frequency pose data is output. A comparison experiment of master–slave trajectory tracking is carried out to evaluate the performance of the proposed method.

2.2 Architecture of Human-Motion Based Teleoperation System

The control flow block diagram of the teleoperation control system designed in this chapter is shown in Fig. 2.1. The whole system can be divided into two parts: human motion capture system and robot pose mapping control system. Based on the multi-node distributed control system architecture of ROS, through the wearable motion capture device PN, the body movement data of the master operator is obtained, and the dual-arm YuMi robot of the pose mapping teleoperation system is controlled to execute the output action command. The live action footage of the slave robot is fed back to the master operator through the cloud camera, so that the operator can perceive the operation status of the remote robot in real time.

2.2.1 Inertia Motion Capture and Pose Data Transformation

The motion capture system consists of a wearable motion capture suit and corresponding software program, which can acquire, process and transmit the wearer's motion signals in real time through a wireless connection. Motion capture suite PN is a versatile and adaptable motion capture device composed of a multi-node Inertial Measurement Unit (IMU). Each IMU consists of a gyroscope, accelerometer, and magnetometer. All IMU nodes are fastened to the operator with elastic straps. Hub is the central processing unit of the wearable mobile capture device PN,

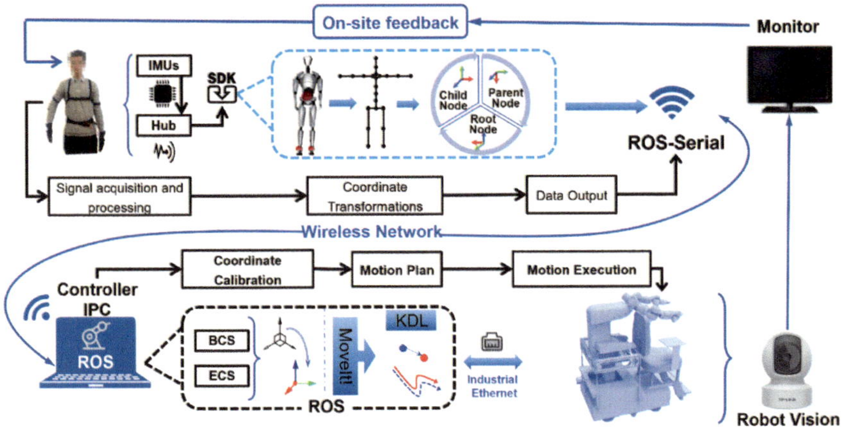

Fig. 2.1 Detailed control block diagram of the proposed telerobotic system, reprinted from Ref. [4], CC BY 4.0

which can transmit and decode the data of each IMU node through Wi-Fi coding and transmit it to the upper computer software Axis Neuron. Axis Neuron is the upper computer software connected to PN device running in Windows operating system, receiving motion data from PN hardware, then processing and exporting the motion data of wearers into a unified BVH (BioVision Hierarchy) format [5] to broadcast and transmit motion data stream. In order to realize the communication between PN motion coding data and robot control host computer, this book realizes the data transmission communication between operator and robot based on ROS_ Serial communication package and Peception Neuron Data Reader SDK.

According to the operator's upper limb movement data obtained from IMU, a human skeleton joint chain model is constructed, as shown in Fig. 2.2. The model consists of hip node, spine 0, spine 1, spine 2, spine 3, shoulder, upper arm, lower arm, and hand node. In the process of motion mapping, the operator's hip node is set as the root node of the human skeleton model, and the hand node is set as the end node of the master operator. Other nodes are matched with other nodes in turn according to their relative motion. The two adjacent nodes are respectively parent node and child node, the child node is the one that further from the hip node.

During the capture of body movements, the inertial motion capture device PN and the upper computer on the robot end communicate with each other via the TCP/ IP protocol based on the ROS_Serial tool package for synchronized data transmission. In this section, 18 IMU nodes are used for motion capture (excluding the hand node, only the limbs and torso nodes are included, among which the upper arm includes three nodes, namely the upper arm, the lower arm and the shoulder), and the maximum real time output frequency of the inertial motion capture system is 120 Hz. The upper computer software Axis Neuron displays the operator's actions in real time through the animated bone model. Due to the difference in body size of different wearers, the initial posture of the animated bone model does not match

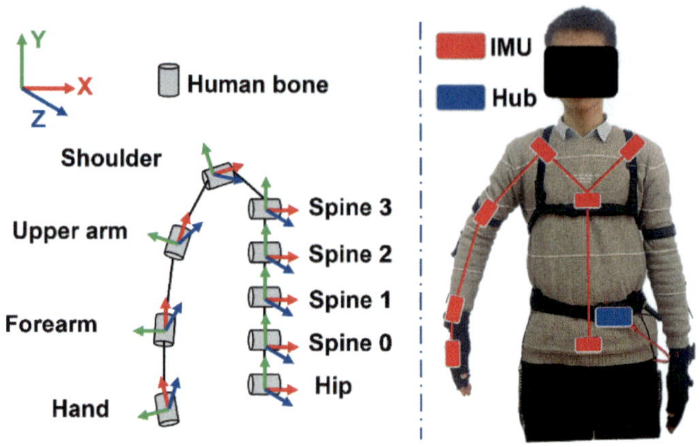

Fig. 2.2 The skeleton model of the upper limb and the IMUs on the body, reprinted from Ref. [4], CC BY 4.0

the posture of human. When the operator wears it for the first time, they can set the parameters of their arm length, shoulder width and height in Axis Neuron, create the corresponding personalized bone model, and execute the motion calibration procedure to synchronize the bone model with the operator's movement. Table 2.1 lists the detailed technical parameters of the PN wearable inertial motion capture device.

The human motion data in BVH format describes the global motion data of each node in the global coordinate system: displacement, velocity and quaternions, and the relative motion data of the child node in the local coordinate system of the parent node: acceleration and angular velocity. In order to realize the transformation of the

Table 2.1 Technical parameter table of wearable inertia motion capture device PN

Device	Parameter	Parameter value
IMU node of the PN	Dimension	12.5 mm × 13.1 mm × 4.3 mm
	Wireless connection range	360°
	Accelerometer range	± 16 g
	Gyroscope range	± 2000 dps
	Minimum resolution	0.02°
	Nodes used in this book	18
Hub processing unit of PN	Dimension	59 mm × 41 mm × 23 mm
	Minimum usable nodes	3
	Maximum number of Connectable nodes	32
	Maximum output frequency	18 nodes:120 FPS 32 nodes:60 FPS

local coordinate system of each skeleton node in the global coordinate system, the position data and attitude data of each child node in the local coordinate system are used to define the transformation relationship with its parent node. The displacement vector and rotation matrix of the child node relative to the parent node are calculated, and the local transformation matrix is obtained as shown in Eqs. (2.1) and (2.2).

$$R = Rot(y, \theta_y)Rot(x, \theta_x)Rot(z, \theta_z) \tag{2.1}$$

$$M_i^{i-1} = \begin{bmatrix} R\,T \\ 0\;1 \end{bmatrix} \tag{2.2}$$

where R represents the rotation matrix of the child node in the local coordinate system of the parent node; $\theta_y, \theta_x, \theta_z$ respectively represents the euler angle along the Y axis, X axis, Z axis; $Rot(y, \theta_y)$, $Rot(x, \theta_x)$, $Rot(z, \theta_z)$ respectively represents the rotation matrix around each single axis of the Y axis, $X\ axis$, Z axis; The vector T represents the displacement vector along the X axis, Y axis, Z axis; M_i^{i-1} represents the local transformation matrix between parent node $i-1$ and child node i.

Taking right arm transformation as an example, coordinate transformation is performed from the operator's hip root node to the hand end node to obtain the overall transformation matrix in the motion chain of the human skeleton model, as shown in Eq. (2.3) below:

$$M_n^0 = \prod_{i=1}^{n} M_i^{i-1} \tag{2.3}$$

where M_n^0 is the pose transformation matrix n represents the Nth node in the Kinematic chain, $n = 0$ represents hip root node.

2.2.2 *Components of Single-Arm Robot Teleoperation System*

The operating system of YuMi robot is consistent with that of ABB ordinary six-axis industrial robots. In this book, RobotWare 6.04 system running in IRC5 controller is used to control the twin robot arms of YuMi robot. The single arm of YuMi robot is composed of redundant seven-axis robot arm and end actuator (two-finger servo jig), which are connected to the main control board through industrial Ethernet for independent communication. The detailed technical parameters of YuMi robot are shown in Table 2.2.

In this book, the control frame of robot is designed based on ROS-Industrial. According to the sample-massage protocol, RAPID language is used to program and configure the motion control interface of YuMi robot arm and end effector through Socket communication in the application layer of the robot controller. An Ubuntu industrial computer configured with ROS is used as the upper computer for YuMi robot control, and the hardware layer is connected based on abb-driver function

Table 2.2 Technical parameters of ABB YuMi robot

Device	Parameter entry	Parameter value
YuMi (IRB14000)	Number of robot arms	2
	Single-arm axis count	7
	Single-arm span	0.559 m
	Single-arm Load	0.5 kg
	Protection standard IEC529	IP30
	Power	< 0.17 kW
	End-effector maximum speed	1.5 m/s
	End-effector maximum acceleration	11 m/s^2
	Repeatability positioning accuracy	0.02 mm
	Self-weight	38 kg
Smart gripper (two-finger servo gripper, no vision, no sucker)	Self-weight	0.23 kg
	Maximum load	0.27 kg
	Input voltage	24 VDC
	Maximum power	9 W
	Servo stroke	0–50 mm
	Maximum clamping speed	25 mm/s
	Finger repeated positioning accuracy	0.05 mm
	Maximum holding power	20 N
	Clamping force control accuracy	3 N

package. Motion planning and control of YuMi robot arms are completed in MoveIt based on the YuMi open source URDF model provided by Krug et al. [6]. Since YuMi's robot arms and end actuator are independently controlled by Socket communication according to different network addresses, this chapter divide YuMi robot into four independent motion planning groups: left arm, right arm, left hand and right hand. The base coordinate system of YuMi robot is located at the center point of the front edge of the base, so as to facilitate the unified coordinate transformation of the left and right arms.

There are two ways to solve the robot's target state based on URDF model by using various kinematic solvers in MoveIt: (1) Given the target joint space directly, meaning the robot interpolates from its current joint space to the target joint space during the path planning process by specifying the angular rotation of each joint axis; (2) Given the target end-effector position and orientation, the controller needs to first inverse kinematics to determine the joint space corresponding to the end-effector target pose. Then, based on the inverse kinematics solution, interpolate from the initial state to the target state in the joint space, thereby outputting temporal joint control data. The two motion planning forms mentioned above determine two different human–robot action mapping rules: direct mapping in joint space and point-to-point mapping at the end effector, as illustrated in Fig. 2.3. Due to the differences between the operator's

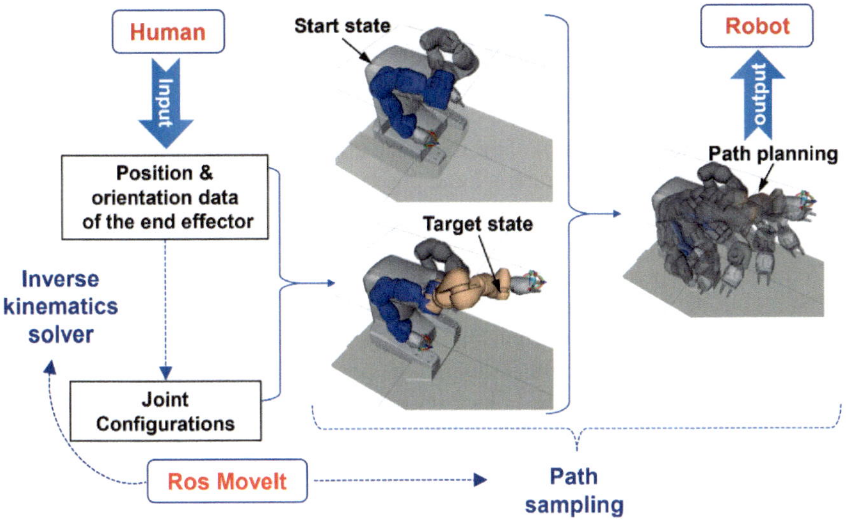

Fig. 2.3 Two types of motion planning methods in ROS moveit

joint configuration and the robot's joint configuration, the master end cannot directly obtain continuous joint data output corresponding to the slave robot's configuration. Therefore, this section adopts a point-to-point mapping at the end effector and inverse kinematics at the slave end to achieve real-time human–robot pose mapping for teleoperation.

Based on the previously implemented human body pose transformations captured by the operator, the global transformation matrix of the operator's end-effector hand node relative to the hip node can be obtained. The data from the wearable device is connected to the industrial control computer on the robot side via ROS Serial, and the real-time pose transformation data is sent, after the ROS TF coordinate transformation [7] the pose data of the operator's hand node in the body root node coordinate system can be obtained. Subsequently, based on the kinematic model of the robot side, the target joint space can be solved inversely to output real-time temporal joint data for controlling the slave-end manipulator. The real-time transformation matrix of the operator's hand coordinate system in the world coordinate system can be represented by the following Eq. (2.4).

$$M_h = \begin{bmatrix} R_h & T_h \\ 0 & 1 \end{bmatrix} = M_n^0 \times M_0^{global} \tag{2.4}$$

where M_0^{global} is the transformation matrix for the operator's hip root node relative to the origin of the world coordinate system, R_h and T_h respectively respresent the rotation matrix and translation vector of the operator's hand coordinate system relative to the world coordinate system.

The operator's hand position p_h is obtained by solving for T_h, and the operator's hand orientation p_r is obtained by solving for R_h. As the representation of the robot arm's end-effector orientation use Euler angles, deadlocks are prone to occur during the solving process. To ensure the continuity of pose mapping, in this section of teleoperation, the orientation q_r of the robot end effector is kept consistently oriented, during teleoperation, the operator's hand pose is directly mapped to the robot end, with the operator's hand pose q_h remaining relatively stable, the design of mapping strategies for pose transformations will be explored in Chap. 3 of this book.

To utilize the operator's hand pose data to control the robot's end-effector pose, it is necessary to ensure that their base coordinate systems are aligned. Coordinate transformation is then performed to align the operator's base coordinate system with that of the robot's. Assuming that the relative position between the operator's base coordinate system and the robot's base coordinate system is p_δ, the transformation relationship between the robot's end-effector position p_r and the operator's end-effector position p_h should be as shown in Eq. (2.5).

$$p_r = p_h + p_\delta \tag{2.5}$$

From the input of the operator's motion data to the output of the robot's joint motion data, the inverse kinematic solving for the target joint states consumes a considerable amount of time. Although this point-to-point direct mapping rule can control the robot's end effector to accurately reach the target position in its base coordinate system in one calculation output, it is difficult to ensure the smoothness of continuous action and tracking similarity with the operator's hand trajectory. In order to solve the problem of continuous human–robot end trajectory tracking based on point-to-point direct mapping rules, a single-arm pose mapping method based on local path resample of end trajectory is proposed in this chapter.

2.3 Single-Arm Motion Mapping Based on Trajectory Resampling

2.3.1 Path-Constrained Trajectory Tracking Algorithm

Based on the point-to-point end-effector mapping method, this section proposes a new method for human–robot pose mapping based on local path resampling of end-effector motion trajectories. By periodically resampling the local trajectory of the operator's hand, it constrains the motion path of the slave robot, making it more suitable for trajectory tracking based on continuous motion of the operator's hand. The specific process of this method is described as follows: At predefined intervals ΔT, the hand pose data of the operator $[p_h(t-\Delta T-\delta t : t), q_h(t-\Delta T-\delta t : t)]$ and the current end effector pose data of the robot $[p_r(t), q_r(t)]$ are input as constraint points for the robot's desired motion path planning, where δt is the refresh time for path planning, including the time taken for kinematic solving and the transmission of motion data. Within the next local path sampling time ΔT, the point-to-point end-effector pose data corresponding to the local path trajectory of the master operator $[p_h(t+\delta t : t+\Delta T+\delta t), q_h(t+\delta t : t+\Delta T+\delta t)]$ is taken as the expected end-effector motion data for the robot's next sampling cycle, and output to the robot controller for execution. The pseudocode implementation of the local path resampling mapping method is shown in Algorithm 2.1.

Algorithm 2.1 : Local path resampling process of man-machine trajectory mapping algorithm

Input: $[p_h(t-\Delta T-\delta t:t), q_h(t-\Delta T-\delta t:t)]$, $[p_r(t), q_r(t)]$
Constraint conditions : $q_\delta = 0$
Output: $[p_r(t+\delta t:t+\Delta T+\delta t), q_r(t+\delta t:t+\Delta T+\delta t)]$
while (The program was not forcibly stopped) do
  (I) Mapping human posture data
  $p_r(t-\Delta T-\delta t:t) \leftarrow p_h(t-\Delta T-\delta t:t)+p_\delta$
  $q_r(t-\Delta T-\delta t:t) \leftarrow q_h(t-\Delta T-\delta t:t)$
  (II) Local path resampling
  Waypoints $\leftarrow$ ($[p_r(t), q_r(t)]$)
  for $i = t-\Delta T-\delta t:t$
    Waypoints $\leftarrow$ Add (Waypoints, $[p_h(i), q_h(i)]$)
  end
  (III) Path planning
  Traj $\leftarrow$ MoveitPlan (Waypoints)
  (IV) Robot control pose output
  for $i = t+\delta t:t+\Delta T+\delta t$
    $[p_r(i), q_r(i)] \leftarrow$ Resample(Traj)
  end
  (V) Clear the waypoint cache
  Clear Waypoints
end

Figure 2.4 shows the cyclic execution process of the constrained human robot pose mapping method based on local path resampling. Setting the local sampling interval ΔT for the operator's continuous trajectory, δt represents the refresh time for the robot's path planning, simultaneously, setting the actual execution time ΔT for the robot's movements, which implies that the robot's end-effector must move faster than the operator's hand. Under the rule of looping through re-sampling local trajectories, the robot's actual execution time will lag the operator's actual movements by $\Delta T + 2\delta t$. The workflow of the algorithm can be flexibly adjusted in practical use, allowing for a time interval between each sampling cycle of the operator's hand trajectory. Simultaneously, it is possible to optimize the configuration of values ΔT and δt during the mapping process based on different operators' motion speeds, and δt is related to the actual motion performance of the robot used. The value of ΔT is achieved in subsequent experiments by changing the resampling frequency F_{re}, and their relationship is represented by the following Eq. (2.6).

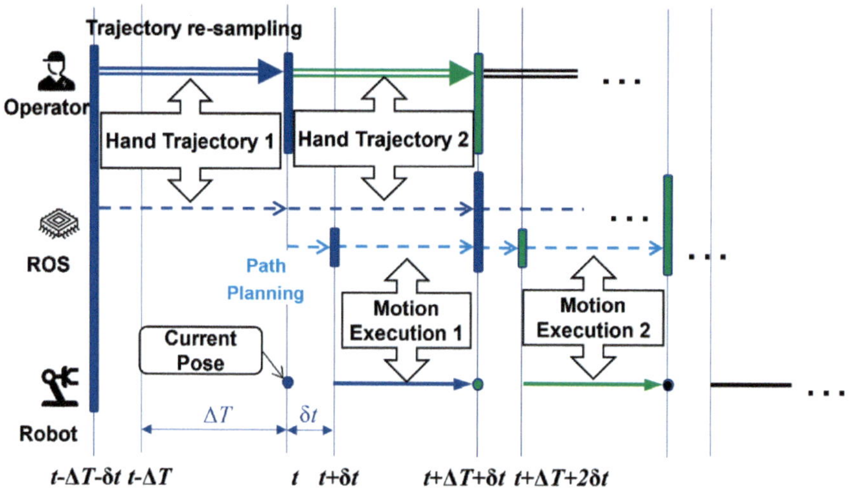

Fig. 2.4 Motion mapping execution process based on trajectory resampling

$$F_{re} = \frac{1}{\Delta T} \tag{2.6}$$

2.3.2 Human–Robot Motion Tracking Performance Evaluation

To verify the master–slave trajectory tracking performance of the human–robot pose mapping method based on local path resampling constraints, this section designs end-effector trajectory tracking experiments with two levels of complexity. Here uses simple spatial trajectories such as horizontal lines "—", vertical lines "|", triangles "△", and arcs "⌒", and chose the spatial writing patterns of the letters "Z", "J", and "U" (representing the abbreviation for Zhejiang University) as complex trajectories. To explore the impact of local path re-sampling frequency on end-effector trajectory tracking performance, the above two groups of trajectories are performed by the operator under five different resampling frequencies (F_{re} = 2.0 Hz, 3.5 Hz, 5 Hz, 7.5 Hz, 10.0 Hz) during the pose mapping teleoperation. Due to the discrepancy in end-effector velocities between the robot and the operator, there inevitably exists a delay between the end-effector motion trajectory of the robot and the hand motion trajectory of the operator during teleoperation. To objectively assess the similarity of end-effector motion trajectories between the robot and the operator, the study applies the Dynamic Time Warping (DTW) algorithm to reduce temporal distortion in trajectory signals [8], and computes the euclidean distance between corresponding spatial trajectory curves (the following content uses the DTW distance representation) to quantify the similarity between trajectories.

2.3.2.1 Evaluation Experiment of Basic Trajectory Tracking Performance

In the experiment of simple spatial trajectories tracking, operators draw four simple trajectories "—", "|", "△", "⌒" respectively, during a series of teleoperation experiments. In the robot's upper computer program, the proposed local path resampling constraint rules and master–slave pose mapping method are deployed, the end effector of the robot executes motion paths according to the trajectory of the operator's hand, to ensure the reliability of performance evaluation data, each set of trajectory tracking experiments is respectively repeated five times at five different resampling frequencies. Figure 2.5 provides a visual representation of the master–slave end tracking performance for the four simple trajectories at $F_{re} = 7.5$ Hz.

The tracking signals of the four simple trajectories at the 7.5 Hz resampling frequency, as shown in Fig. 2.5, using DTW to temporally align raw positional data signals, the results are depicted in Figs. 2.6, 2.7 and 2.8.

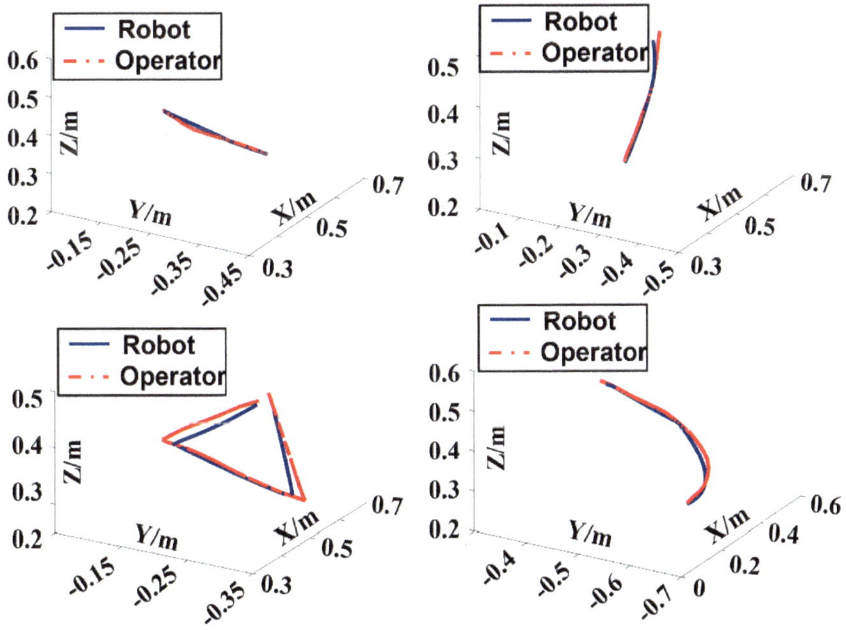

Fig. 2.5 The trajectory tracking performance of four simple trajectories at $F_{re} = 7.5$ Hz, reprinted from Ref. [4], CC BY 4.0

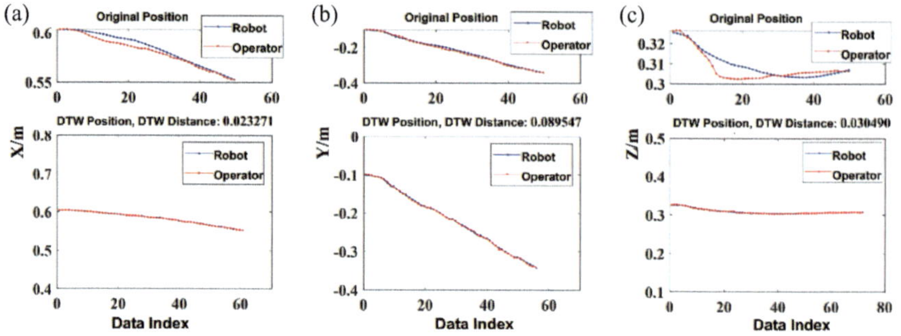

Fig. 2.6 Raw and DTW aligned position data of the trajectory "—". **a** X-axis position. **b** Y-axis position. **c** Z-axis position

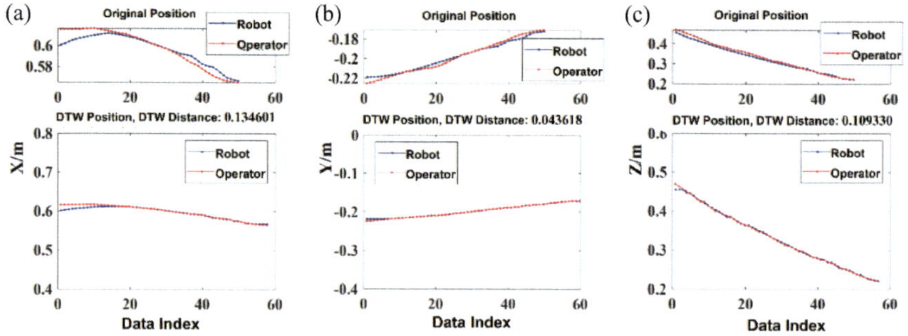

Fig. 2.7 Raw and DTW aligned position data of the trajectory "|". **a** X-axis position. **b** Y-axis position. **c** Z-axis position

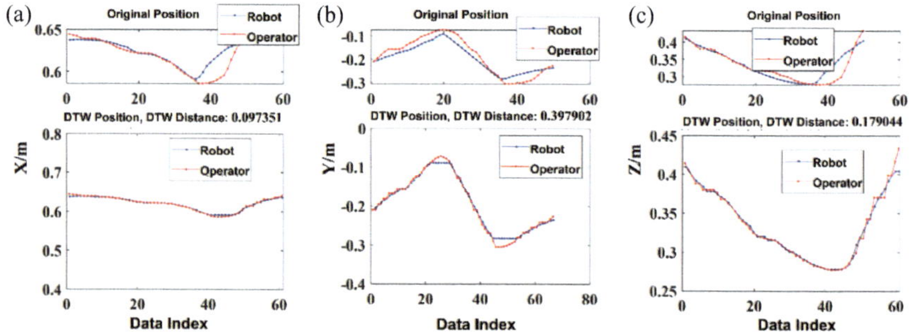

Fig. 2.8 Raw and DTW Aligned position data of the trajectory "△". **a** X-axis position data. **b** Y-axis position data. **c** Z-axis position data

Figure 2.6 shows the position signal data along the XYZ axes during the "—" trajectory mapping process. It can be observed that during the creation of this trajectory, the position changes significantly along the Y-axis. Due to the differences in motion characteristics between the operator and the robot, the position signals along the X and Z axes are not aligned. Therefore, using DTW for time-domain correction results in better alignment of the signals, facilitating further analysis and post-processing.

Figure 2.7 shows the position signal data along the XYZ axes during the "|" trajectory mapping process. It can be observed that, unlike the "—" trajectory mapping shown in Fig. 2.6, the position changes significantly along the Z-axis during the creation of this trajectory. Furthermore, DTW time-domain correction is applied to the position signals along the X and Y axes to achieve time-domain alignment.

Figure 2.8 shows a comparative chart of the position signal data for the operator's hand forming the "Δ" trajectory. Compared to the simpler "|" and "—" trajectories previously discussed, the "Δ" trajectory primarily exhibits positional changes along the Y and Z axes. During the trajectory formation, the operator remains mostly within a spatial plane perpendicular to the X axis, hence changes in the X-axis position are not significant. The comparison shows that the "Δ" trajectory has more unaligned parts in the original signals compared to the "|" and "—" trajectories, thus the effects of DTW (Dynamic Time Warping) time-domain correction are more pronounced.

Figure 2.9 shows the end position data changes during the operator's formation of the "⌒" trajectory. Unlike the previous "|", "—", and "Δ" trajectories, the "⌒" trajectory exhibits varying degrees of changes along the X, Y, and Z axes. The comparison of the original signals with those aligned after DTW time-domain correction is shown in the figure.

After obtaining the trajectory data, the DTW distance between the operator's and robot's end trajectories is calculated using the DTW algorithm. To more comprehensively assess the similarity of human–robot end positions and postures at trajectory path points, this section calculates the root mean square values of position data P_{RMS} and posture data Q_{RMS} as the evaluation metrics for overall curve position tracking

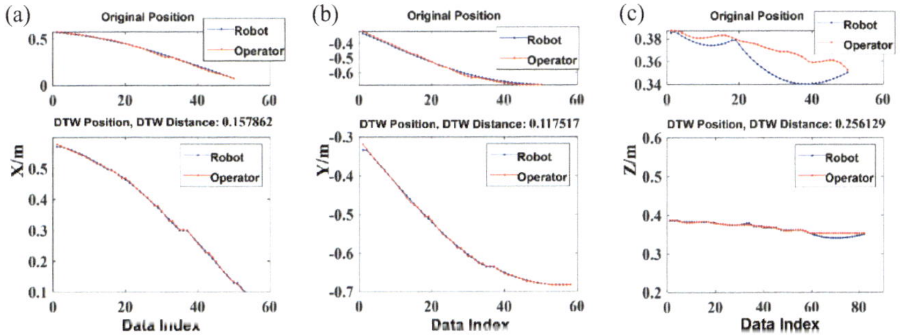

Fig. 2.9 Raw and DTW aligned position data of the trajectory "⌒". **a** X-axis position data. **b** Y-axis position data. **c** Z-axis position data

error and posture tracking error, respectively. The calculation methods are shown in Eqs. (2.7) and (2.8).

$$P_{\mathrm{RMS}} = \sqrt{(x^2 + y^2 + z^2)/3} \tag{2.7}$$

$$Q_{\mathrm{RMS}} = \sqrt{(r^2 + p^2 + y^2)/3} \tag{2.8}$$

Table 2.3 displays the DTW distances between the position data (x, y, z) and posture data (r, p, y) of two trajectories, one from an operator and the other from a robot, taking the mean of five repeated experiments for each trajectory. Statistical analysis data is illustrated in Fig 2.10a and b. From the statistical analysis results in Table 2.3 and Fig. 2.10, it can be observed that the DTW distances decrease as the trajectory resampling frequency increases, indicating an improving similarity in position and posture between human–robot end trajectories. At 7.5 Hz and 10.0 Hz, the DTW distances do not exceed 2.00 mm, but the distance at 10.0 Hz is 1.80 mm, which is higher than the 1.05 mm at 7.5 Hz. This is due to the higher number of sampling points per unit time at 10.0 Hz, making motion planning more challenging and reducing the similarity of the final movement paths. Overall, the results from the simple spatial trajectory tracking experiment suggest that the best tracking performance ($P_{\mathrm{RMS}} = 1.05$ mm) can be achieved at $F_{re} = 7.5$ Hz.

2.3.2.2 Complex Trajectory Tracking Performance Evaluation Experiment

In the experiment with complex hand trajectories, the operator creates the writing trajectories of the three letters "Z", "J", and "U" in space, as illustrated in Fig. 2.11. To ensure data reliability, this section conducts the complex letter writing trajectory tracking experiment five times at five different resampling frequencies. Figure 2.11 displays the remote operation trajectory tracking performance of the three complex trajectories at $F_{re} = 7.5$ Hz during the experiment. Similar to the simple trajectory tracking experiment, the DTW algorithm is used to compare the remotely operated mapping trajectory data obtained from human–robot interaction. During this process, the data from five repeated experiments for each trajectory at each frequency are averaged, as shown in Table 2.4.

Furthermore, an overview of data from Tables 2.3 and 2.4 indicates that the DTW distances for complex trajectories are significantly greater than those for simple trajectories. Similar to the simple trajectory experiment, the root mean squares of position and posture data are calculated, as shown in Fig. 2.12a and b, with the positional tracking deviations for operators and robots in complex trajectories remaining within 11.00 mm. From the results of experiments at sampling frequencies above 3.5 Hz, a similar conclusion to that of the simple trajectory experiments can be drawn: the higher the resampling frequency, the better the robot's tracking performance, also achieving the best tracking performance for complex trajectories at $F_{re} = 7.5$ Hz

Table 2.3 DTW distances for human–robot motion tracking for simple trajectories

F_{re} (Hz)	Trajectory	Position data (mm)				Posture data (rad)			
		x	y	z	P_{RMS}	r	p	y	Q_{RMS}
2.0	—	28.71	0.04	17.35	19.37	0.11	1.27	0.36	0.76
	∣	2.80	0.82	7.25	4.51	0.04	0.00	0.00	0.00
	△	2.79	18.63	44.10	27.69	0.11	0.04	0.06	0.04
	⌒	66.90	12.22	1.03	39.27	0.00	0.00	0.24	0.15
	Mean	25.30	7.93	17.43	18.32	0.06	0.33	0.17	0.21
3.5	—	0.50	0.32	0.35	0.40	0.00	0.02	0.00	0.01
	∣	0.12	0.00	0.10	0.09	0.00	0.01	0.00	0.00
	△	13.62	65.71	5.74	38.88	0.04	0.24	0.39	0.26
	⌒	0.70	1.32	0.03	0.86	0.00	0.00	0.00	0.00
	Mean	3.74	16.84	1.55	10.00	0.01	0.07	0.10	0.07
5.0	—	0.01	0.16	0.01	0.09	0.00	0.00	0.00	0.00
	∣	0.14	0.17	0.31	0.22	0.00	0.02	0.02	0.01
	△	1.06	30.00	3.18	17.43	0.01	0.00	0.11	0.06
	⌒	11.39	0.61	4.35	7.05	0.01	0.17	0.02	0.10
	Mean	3.15	7.73	1.96	4.95	0.01	0.05	0.04	0.03
7.5	—	0.01	0.21	0.02	0.12	0.00	0.00	0.00	0.00
	∣	1.09	0.12	0.45	0.69	0.00	0.01	0.01	0.01
	△	0.31	4.60	1.73	2.84	0.00	0.01	0.03	0.03
	⌒	0.64	0.52	2.11	1.31	0.01	0.00	0.00	0.01
	Mean	0.51	1.36	1.08	1.05	0.00	0.01	0.01	0.01
10.0	—	0.01	0.49	0.05	0.29	0.00	0.00	0.00	0.00
	∣	0.01	0.01	0.11	0.06	0.00	0.00	0.00	0.00
	△	0.40	10.26	1.15	5.97	0.00	0.00	0.03	0.03
	⌒	2.40	1.07	1.32	1.70	0.00	0.00	0.01	0.02
	Mean	0.71	2.96	0.66	1.80	0.00	0.00	0.01	0.01

($P_{RMS} = 5.10$ mm). Since there is almost no change in the operator's hand orientaion during the drawing of the three complex trajectories, a clear representation of orientaion tracking performance cannot be derived from Table 2.4.

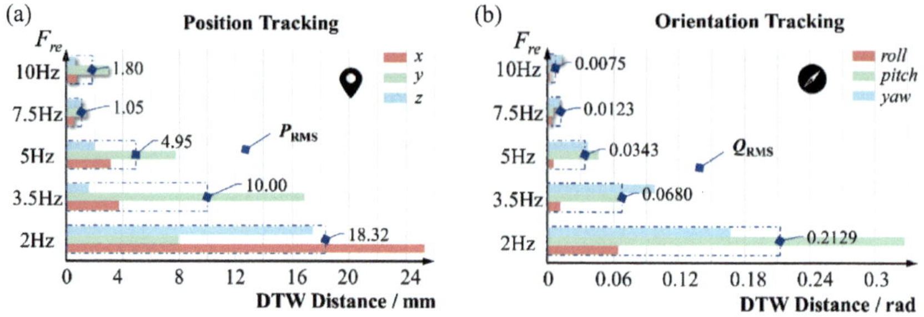

Fig. 2.10 Statistical analysis of tracking performance for simple trajectories, reprinted from Ref. [4], CC BY 4.0. **a** Position tracking accuracy. **b** Posture tracking accuracy

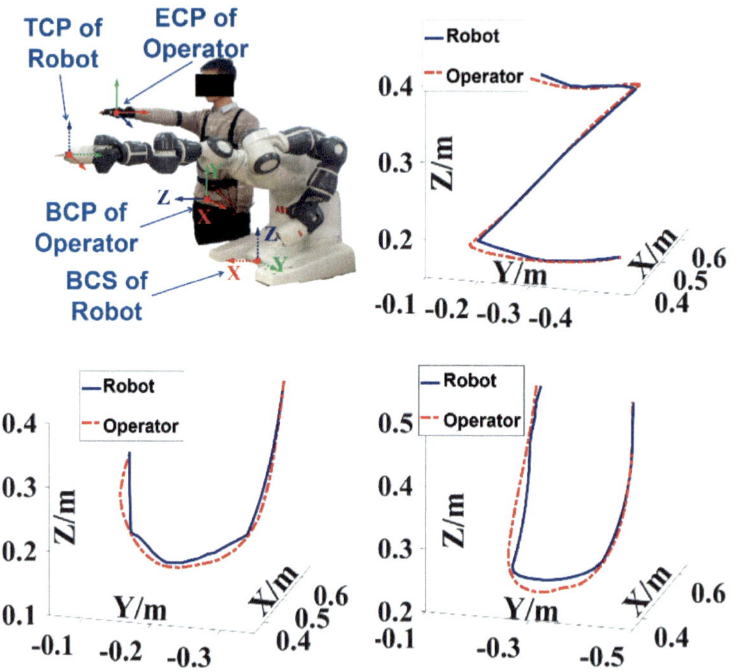

Fig. 2.11 Tracking performance of three complex trajectories at $F_{re} = 7.5\,\text{Hz}$

2.4 Summary

Capturing the limb motion postures of an operator to map and control a robot is a convenient and intuitive method of teleoperation. This chapter focuses on the YuMi dual-arm collaborative robot as the teleoperation slave control entity, using the wearable inertial motion capture device PN to capture the master operator's limb motion

Table 2.4 DTW distances for human–robot motion tracking for complex trajectories

F_{re} (Hz)	Trajectory	Position (mm)				Orientaion (rad)			
		x	y	z	P_{RMS}	r	p	y	Q_{RMS}
2.0	Z	2.45	26.78	2.71	15.60	0.01	0.07	0.09	0.06
	J	0.13	1.35	8.95	5.23	0.03	0.00	0.01	0.01
	U	4.59	0.46	4.16	3.59	0.01	0.05	0.02	0.03
	Mean	2.39	9.53	5.27	6.44	0.02	0.04	0.04	0.03
3.5	Z	0.33	52.07	0.24	30.07	0.01	0.07	0.15	0.09
	J	0.94	0.27	9.36	5.43	0.07	0.00	0.09	0.05
	U	0.99	2.95	4.56	3.18	0.03	0.00	0.02	0.01
	Mean	0.75	18.43	4.72	10.99	0.04	0.02	0.09	0.05
5.0	Z	7.08	28.45	7.44	17.47	0.05	0.23	0.09	0.14
	J	1.71	1.66	7.08	4.31	0.05	0.01	0.01	0.01
	U	4.94	1.37	3.41	3.56	0.02	0.14	0.10	0.10
	Mean	4.58	10.49	5.98	7.46	0.04	0.13	0.07	0.08
7.5	Z	2.00	16.99	1.79	9.93	0.01	0.08	0.05	0.10
	J	1.47	1.11	2.54	1.81	0.03	0.03	0.02	0.05
	U	5.80	2.97	8.87	6.35	0.05	0.07	0.03	0.09
	Mean	3.09	7.02	4.40	5.10	0.03	0.06	0.04	0.08
10.0	Z	0.56	17.71	1.29	10.26	0.02	0.01	0.10	0.10
	J	1.67	0.68	38.55	22.28	0.13	0.06	0.01	0.14
	U	0.96	1.20	5.17	3.11	0.02	0.01	0.04	0.05
	Mean	1.07	6.53	15.00	9.47	0.05	0.03	0.05	0.05

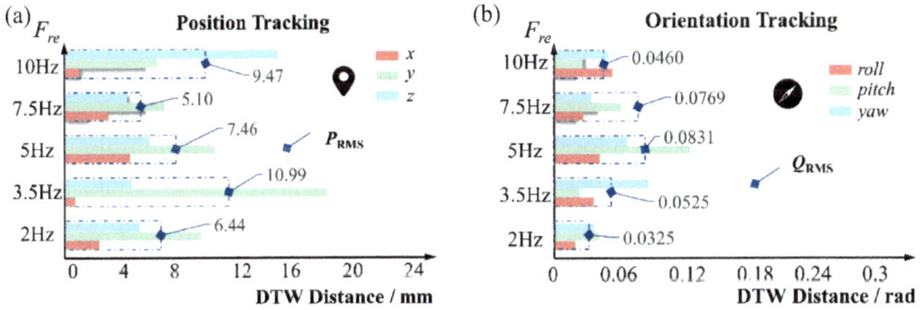

Fig. 2.12 Statistical analysis of tracking performance for complex trajectories. **a** Position tracking accuracy. **b** Orientation tracking accuracy

data, and establishes a human skeleton model motion chain through coordinate transformation. A distributed communication architecture design between the YuMi robot and the wearable device was implemented based on ROS, achieving point-to-point end pose mapping control. A method based on local path resampling constraints for human–robot motion mapping is proposed, which can accomplish the slave robot's mapping and tracking of the master operator's continuous hand trajectory. Comparative experiments were also designed to evaluate trajectory tracking performance, identifying the optimal resampling frequency for the best trajectory tracking results. The main conclusions of this chapter are as follows:

(1) A teleoperation system for mapping the upper limb movements of the master operator to control the pose of the slave robot has been designed and implemented, enhancing the convenience and intuitiveness of the operator's remote control of the robot. By performing coordinate transformations on the skeletal model data obtained from motion capture, the overall transformation matrix in the human skeleton model motion chain is derived, thereby obtaining the operator's hand pose data in the global coordinate system. Based on ROS, real-time data transmission of the operator's end pose data to the slave robot's controller has been implemented, laying the foundation for the development of human–robot pose mapping teleoperation control algorithms and verification of control performance.
(2) Based on the operator's end pose mapping to the slave robot's desired end pose, individually solving and outputting control commands cannot achieve high-frequency pose mapping, leading to uneven speeds in the robot's end effector movement and jerky, unsmooth tracking paths. This chapter introduces a human–robot motion mapping algorithm based on local path resampling constraints. In the mapping process, the operator's upper limb hand pose data and the current robot's end effector pose data are planned through path resampling. This approach improves the tracking accuracy and similarity of the master–slave trajectories during low-frequency pose mapping outputs, achieving continuous trajectory tracking and mapping between the operator and the robot.。
(3) To validate the proposed local path resampling constraint algorithm for continuous trajectory tracking performance in pose mapping teleoperation, performance tests are conducted on four simple trajectories and three complex spatial trajectories at five different resampling frequencies. The euclidean distances between corresponding spatial curves are calculated to quantitatively assess the similarity of the trajectories. The results indicate that both simple and complex trajectories achieve the best tracking performance at $F_{re} = 7.5\,\text{Hz}$. For simple motion trajectories such as straight lines and complex motion trajectories such as spatial characters, the human–robot end-trajectory tracking errors are reduced to position tracking error performances of $P_{\text{RMS}} = 1.05$ mm and $P_{\text{RMS}} = 5.10$ mm, respectively.

References

1. Zahavi A, Haeri SN, Liyanage DC et al (2020) A dual-arm robot for collaborative vision-based object classification. In: Proceedings of the Biennial Baltic electronics conference, Tallinn, Estonia, pp 1–5
2. Choo CZY, Chow JY, Komar J (2022) Validation of the perception neuron system for full-body motion capture. PLoS ONE 17(1):e0262730
3. Quigley M, Conley K, Gerkey B et al (2009) ROS: An open-source robot operating system. In: Proceedings of the ICRA workshop on open source software, Kobe, Japan, p 5
4. Lyu H, Yang G, Zhou H et al (2020) Teleoperation of collaborative robot for remote dementia care in home environments. IEEE J Transl Eng Health Med 8:1400510
5. Meredith M, Maddock S (2001) Motion capture file formats explained. Department of Computer Science, University of Sheffield 211:241–244
6. Makris S, Tsarouchi P, Surdilovic D et al (2014) Intuitive dual arm robot programming for assembly operations. CIRP Ann Manuf Technol 63(1):13–16
7. Foote T (2013) TF: the transform library. In: Proceedings of the IEEE conference on technologies for practical robot applications, Woburn, USA, pp 1–6
8. Wang Y, Lei P, Zhou H et al (2014) Using DTW to measure trajectory distance in grid space. In: Proceedings of the IEEE international conference on information science and technology, Shenzhen, China, pp 152–155

Chapter 3
Bimanual Human-Motion Based Robot Teleoperation

Abstract This chapter introduces a dual-arm pose mapping teleoperation method that addresses the control challenges associated with robotic dual-arm coordination. Unlike single-arm systems, dual-arm robots can execute more complex tasks, but controlling both arms simultaneously can be burdensome for the operator. Traditional pose mapping systems struggle with coordinating actions between the arms and often require additional input devices for linked control. Building on single-arm teleoperation techniques, this chapter proposes a flexible and intuitive dual-arm pose mapping approach that allows seamless switching between operator control and fixture control while improving task efficiency. A significant improvement in this method is the incremental pose mapping strategy, which uses real-time relative human–robot pose transformation. This eliminates the need for tedious absolute coordinate calibration at the start of each task, simplifying the setup process and enhancing flexibility during operation. To validate the effectiveness of this approach, a pick-and-place task experiment was conducted, comparing the proposed incremental mapping method with traditional direct mapping teleoperation. Results indicate that the new method improves operational efficiency and is more accessible for users of varying skill levels.

Keywords Dual-arm motion mapping · Human-motion based teleoperation · Incremental motion mapping · Hand gesture · Hybrid motion mapping

3.1 Introduction

Dual-arm robots, compared to single-arm robots, can perform more flexible operations and are capable of executing more complex tasks through coordinated dual-arm actions. However, in pose mapping teleoperation systems that control robotic operations, the master operator must control the movement of the mechanical arms to position the end effector at the target location and manipulate the end effector to perform gripping actions. This often requires frequent switching between control objects and coordinating both arms to accomplish tasks, presenting a substantial

H. Lyu et al., *Human Motion Awareness and Robot Teleoperation*, Advanced Topics in Science and Technology in China 4, https://doi.org/10.1007/978-981-96-6545-7_3

control burden. Ordinary pose mapping strategies struggle to coordinate the relationship between different actions and generally require the use of additional input devices for linked control. Building on the single-arm pose mapping teleoperation presented in Chap. 2, this chapter designs a dual-arm pose mapping teleoperation method that not only allows flexible switching between operator control and fixture control but also supports intuitive mapping control of dual-arm poses. Traditional pose mapping teleoperation systems require tedious master–slave absolute coordinate calibration at the start of each mapping, affecting operational efficiency. This chapter proposes an incremental pose mapping teleoperation method based on real-time relative human–robot pose transformation, which not only enhances flexibility during the mapping process but also simplifies the pre-calibration process in human–robot pose mapping teleoperation. Finally, this chapter sets up a pick-and-place task experiment to compare and evaluate the performance of the proposed incremental pose mapping teleoperation against direct mapping teleoperation among users of different skill levels.

3.2 Hybrid Mapping Technique of Hand Gesture and Upper-Limb Motion

In this section, to achieve intuitive teleoperation of robots by operators, a hybrid human–robot motion mapping teleoperation method based on the coupling of operator hand gestures and limb movements, GuLiM (Gesture-Upper Limb Motion Mapping), is proposed, as shown in Fig. 3.1. Compared to Chap. 2, the wearable inertial motion capture device PN used in this chapter consists of 32 IMU nodes, including two gloves that capture finger movements, with 7 IMU nodes per hand. The operator's upper limb movements and hand gestures are captured using the wearable device and transmitted to the robot end through the designed kinematic mapping strategy to control the robot's arms and end effector. The proposed gesture-mixed mapping teleoperation method introduces incremental pose mapping compared to traditional direct mapping teleoperation controls. It allows operators to more flexibly intervene in the action transfer process through predefined hand gestures while transferring the operator's upper limb actions to the robot, thus achieving coordinated control over various operational objects during the operator's dual-arm pose mapping teleoperation process.

In the actual process of pose mapping teleoperation, it is crucial to accurately convey the operator's intentions and commands for controlling the movement of the robot arm and the actions of the end effector. Specifically, operators need to be able to autonomously and intuitively decide when to control the movement of the robot arm and when to control the movement of the end effector, without interference between the two controls. Therefore, this section predefines various hand gesture rules, as shown in Fig. 3.2, for enabling control of the robot arm and the control of the end effector.

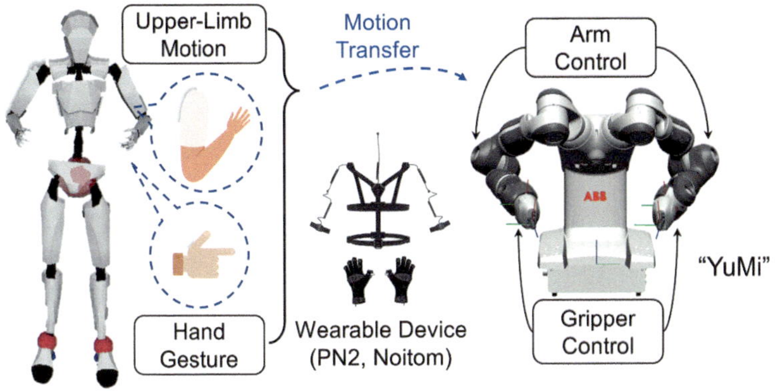

Fig. 3.1 Teleoperation architecture based on hand gesture and upper-limb motion, reprinted from Lyu et al. [1], Copyright (2025), with permission from IEEE

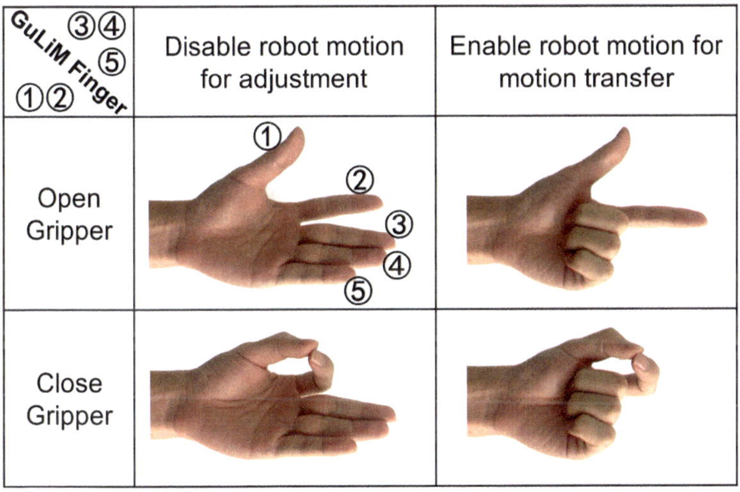

Fig. 3.2 Hand gesture rule definition for hybrid GuLiM mapping technique, reprinted from Lyu et al. [1], Copyright (2025), with permission from IEEE

To better adapt to human motion habits during object grasping, the following rules are set: (1) The bending state of the index finger is used to control the gripper, determined by changes in the posture of the fingertip coordinate system relative to the hand coordinate system $\{F_{\text{handR}}\}$ (using the z-component z_{ind} of quaternion $q_{\text{ind}} = (x_{\text{ind}}, y_{\text{ind}}, z_{\text{ind}}, w_{\text{ind}})$ as the criterion); (2) The bending state of the middle finger is used to enable or disable motion control and teleoperation action transfer of the robot arm, also determined by changes in the posture between the middle fingertip coordinate system and the hand coordinate system $\{F_{\text{handR}}\}$ (using the z-component z_{mid} of quaternion $q_{\text{mid}} = (x_{\text{mid}}, y_{\text{mid}}, z_{\text{mid}}, w_{\text{mid}})$ as the criterion). In this section, the trigger threshold for gripper control is set at $Tz_{\text{ind}} = 0.2$, and the trigger threshold for robotic arm control is set at $Tz_{\text{mid}} = 0.5$. The aforementioned gesture definitions enhance the coordination and convenience of motion transfer between the operator and the robot.

3.2.1 Incremental Motion Mapping Strategy

Based on the defined gesture rules, the motion mapping issue between the operator and the robot can be described as follows: Given the current motion state of the operator (the process amount from the start state to the current state), the robot executes corresponding process actions (starting from a corresponding initial state, moving through the path to reach the respective sub-goals, and ultimately arriving at the designated state). However, due to differences in the kinematic models between humans and robots, it is necessary to define master–slave pose mapping rules to further map the motion capture data from the operator's side into the target position of the robot arm's end effector, as shown in Fig. 3.3.

For position mapping, the hand position data $p_h = (x_h, y_h, z_h)$ of the operator is mapped to the position $p_r = (x_r, y_r, z_r)$ of the end effector. As shown in Fig. 3.3, the initial coordinate system $\{F_{\text{handR}}\}$ of the operator's right hand does not align with the direction of the right gripper's coordinate system. To map the direction of the operator's hand coordinate system to the end effector, a rotation transformation of $\{F_{\text{handR}}\}$ is used to create a new coordinate system $\{F_{\text{handRM}}\}$, which is located at the same position as $\{F_{\text{handR}}\}$ but with a different orientation. Thus, the orientation

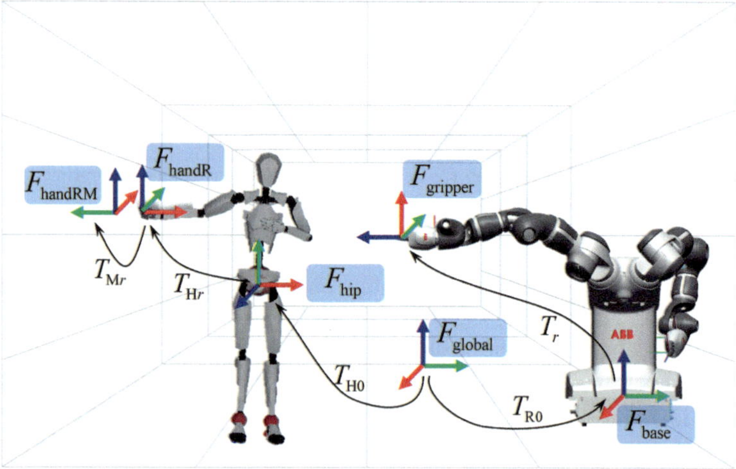

Fig. 3.3 Human–robot coordinate transformation in Cartesian space, reprinted from Lyu et al. [1], Copyright (2025), with permission from IEEE

of $\{F_{\text{handRM}}\}$ can be directly mapped to the robot's right arm end effector $\{F_{\text{gripper}}\}$. Additionally, the default orientation of the coordinate system of the operator's BVH format skeletal local reference system does not align with the traditional Cartesian coordinate system. $\{F_{\text{hip}}\}$ is the local reference system of the BVH skeletal coordinate system, where the Y-axis is perpendicular to the horizontal plane, and the Z-axis and X-axis are perpendicular to each other and lie in the horizontal plane. To address the inconsistency in the definition of coordinate system orientations during the mapping process, this section uses the coordinate data (Z, X, Y) from the BVH format skeletal local coordinate system to map onto (X, Y, Z) in the robot's coordinate system.

The standard DH (Denavit-Hartenberg) model [1] parameters for the single arm of the YuMi robot used in this chapter are shown in Table 3.1. In the table, the parameters θ_j, α_j, a_j, l_j and u_j respectively represent joint angle, link twist angle, link offset, link length, and the lower and upper limits of the joint position, the homogeneous transformation matrix for joint j is derived from Eq. (3.1).

$$
{}^{j-1}_{j}\boldsymbol{T} = \begin{bmatrix} c\theta_j & -c\alpha_j s\theta_j & s\alpha_j s\theta_j & a_j c\theta_j \\ s\theta_j & c\alpha_j c\theta_j & -s\alpha_j c\theta_j & a_j s\theta_j \\ 0 & s\alpha_j & c\alpha_j & d_j \\ 0 & 0 & 0 & 1 \end{bmatrix} \tag{3.1}
$$

where $s\theta_j$ and $c\theta_j$ respectively represent $\sin\theta_j$ and $\cos\theta_j$, $s\alpha_j$ and $c\alpha_j$ respectively represent $\sin\alpha_j$ and $\cos\alpha_j$.

The aforementioned standard D-H model parameters define the forward kinematics of the YuMi robot's single arm from the first joint link to the seventh joint link. Both the left and right arms of the YuMi robot have the same kinematic configuration. However, the base transformation matrices from the YuMi robot's main body base coordinate system $\{F_{\text{base}}\}$ to the base of the first link frame of each arm are different, denoted as ${}^{r_0}_{base}T$ and ${}^{l_0}_{base}T$, respectively. Thus, the forward kinematic transformations for the left and right arms of the YuMi robot can be derived using the following formula:

$$
T_r = {}^{r_0}_{base}T \cdot {}^{0}_{1}T \cdot {}^{1}_{2}T \cdot {}^{2}_{3}T \cdot {}^{3}_{4}T \cdot {}^{4}_{5}T \cdot {}^{5}_{6}T \cdot {}^{6}_{7}T \tag{3.2}
$$

Table 3.1 Standard DH model parameters for single arm of YuMi robot

j	d_j	θ_j	a_j	α_j	l_j	u_j
1	0	θ_1	−30 mm	−90°	−168.5°	168.5°
2	0	θ_2	30 mm	90°	−143.5°	43.5°
3	251.5 mm	θ_3	40.5 mm	−90°	−123.5°	80°
4	0	$\theta_4 - 90°$	40.5 mm	−90°	−290°	290°
5	265 mm	$\theta_5 + 180°$	27 mm	−90°	−88°	138°
6	0	θ_6	−27 mm	90°	−229°	229°
7	36 mm	θ_7	0	0	−168.5°	168.5°

$$T_l = {}^{l_0}_{base}T \cdot {}^{0}_{1}T \cdot {}^{1}_{2}T \cdot {}^{2}_{3}T \cdot {}^{3}_{4}T \cdot {}^{4}_{5}T \cdot {}^{5}_{6}T \cdot {}^{6}_{7}T \tag{3.3}$$

According to data provided in the ABB official manual, the transformation matrix values between the robot base and the first axis of the left and right arms are as follows:

$$ {}^{r_0}_{base}T = \begin{bmatrix} 0.571 & 0.107 & 0.814 & 135 \\ 0.619 & 0.707 & -0.342 & -107 \\ -0.539 & 0.700 & 0.470 & 462 \\ 0 & 0 & 0 & 1 \end{bmatrix} \tag{3.4}$$

$$ {}^{l_0}_{base}T = \begin{bmatrix} 0.571 & 0.107 & 0.814 & 135 \\ -0.619 & 0.707 & 0.342 & 107 \\ -0.539 & -0.699 & 0.470 & 462 \\ 0 & 0 & 0 & 1 \end{bmatrix} \tag{3.5}$$

Based on the kinematic configuration and coordinate transformation analysis mentioned above, to accurately and intuitively map the operator's end actions to the robotic arm's end, this section proposes an incremental pose mapping method, as shown in Fig. 3.4. The operator's hand pose data is typically used to determine the required end-effector pose data for controlling the robot. The poses of the robot's end effector, $pose_r$, and the operator's hand pose, $pose_h$, at moment i are defined as follows:

$$pose_r(i) = (p_r(i), q_r(i)) \tag{3.6}$$

$$pose_h(i) = (p_h(i), q_h(i)) \tag{3.7}$$

where $p_r(i)$, $p_h(i)$ respectively refer to the position vectors of the robot's end effector and the operator's hand, while $q_r(i)$, $q_h(i)$ respectively denote the orientation vectors of the robot's end effector and the operator's hand, represented by unit quaternions.

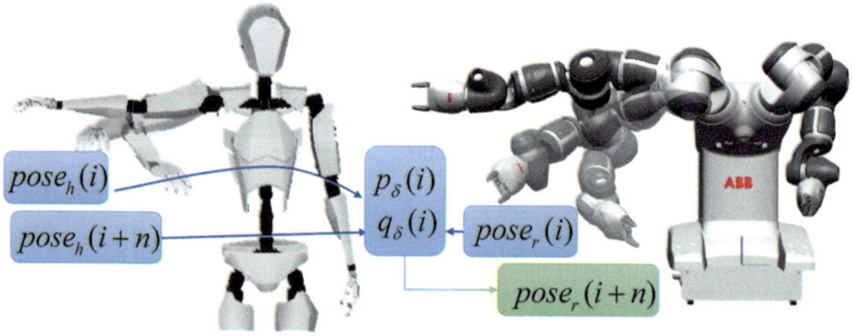

Fig. 3.4 Schematic diagram of incremental pose mapping strategy, reprinted from Lyu et al. [1], Copyright (2025), with permission from IEEE

When the operator starts moving at time i, the relative position $p_\delta(i)$ and the relative orientation $q_\delta(i)$ between the operator and the robot are calculated as follows:

$$p_\delta(i) = p_r(i) - p_h(i) \tag{3.8}$$

$$q_\delta(i) = q_r(i) \otimes q_h(i)^{-1} \tag{3.9}$$

where operator $\otimes$ represents the multiplication of quaternions, and the inverse quaternion $q_h(i)^{-1}$ of the unit quaternion $q_h(i)$ can be obtained by taking the negation of the corresponding components x, y, z of $q_h(i)$.

After obtaining the relative pose at moment i (including the position component $p_\delta(i)$ and the orientation component $q_\delta(i)$), the pose data for robot control at moment $i+n$, as determined by the incremental pose mapping strategy, is defined as follows:

$$p_r(i+n) = p_h(i+n) + p_\delta(i) \tag{3.10}$$

$$q_r(i+n) = q_\delta(i) \otimes q_h(i+n) \tag{3.11}$$

where $p_r(i+n)$ and $q_r(i+n)$ respectively represent the position and orientation of the robot's end effector at moment $i+n$, while $p_h(i+n)$ and $q_h(i+n)$ respectively represent the position and orientation of the operator's hand at moment $i+n$.

With this, the target motion position for the robot end can be obtained. Based on the aforementioned forward kinematics analysis, by inputting the DH parameters into the inverse kinematics solver, the target joint positions can be determined for joint position control on the robot end. In the actual control process of the YuMi robot, this chapter builds on the ROS control framework interface developed in Chap. 2, introducing a high-frequency real-time joint position control interface (EGM, Externally Guided Motion) provided by ABB under option 689-1. The real-time target joint positions obtained from the inverse solution are input into the YuMi robot for position control.

3.2.2 *Dual-Arm Motion Mapping Method with Gesture Recognition*

Based on the proposed incremental pose mapping strategy, operators can transfer their movements to the robot at any time and from any initial pose state, without the need to adjust their own initial poses to fit the robot's starting position. During teleoperation, due to differences in kinematic configurations, the workspaces of the operator and robot are not consistent. Traditional mapping methods adapt workspaces by changing the mapping scale parameters between different operators and robots, often involving tedious calibration work. The new hybrid mapping method proposed in this study, based on operator gestures and limb movements, resolves this issue by eliminating the need for calibration steps, significantly reducing the learning cost and operational difficulty for operators, and achieving flexible mapping of actions between operators and robots.

Algorithm3.1 : GuLiM master-slave mapping algorithm logic and process steps.

Input: Hand gesture and upper-limb motion data of operator

Output: Joint configurations for robot arm control, gripper state

while (termination condition not satisfied) do

  (i) get current state of operator and robot

  $q_{\text{mid}} \leftarrow (x_{\text{mid}}, y_{\text{mid}}, z_{\text{mid}}, w_{\text{mid}})$

  $q_{\text{ind}} \leftarrow (x_{\text{ind}}, y_{\text{ind}}, z_{\text{ind}}, w_{\text{ind}})$

  $pose_r(i) \leftarrow (p_r(i), q_r(i))$

  $pose_h(i) \leftarrow (p_h(i), q_h(i))$

  Parallel procedure A

    (ii) update the relative pose

    if ($z_{\text{mid}} < Tz_{\text{mid}}$) then

      $p_\delta(i) \leftarrow p_r(i) - p_h(i)$

      $pose_h(i) \leftarrow (p_h(i), q_h(i))$

      Enable Robot Arm Motion

    else

      Disable Robot Arm Motion

    (iii) generate robot target pose

    $p_r(i+n) \leftarrow p_h(i+n) + p_\delta(i)$

    $q_r(i+n) \leftarrow q_\delta(i) \otimes q_h(i+n)$

    $pose_r(i+n) \leftarrow (p_r(i+n), q_r(i+n))$

    (iv) inverse kinematic solution

    Joint configuration $\leftarrow$ *trac_ik*($pose_r(i+n)$)

  Parallel procedure B

    (v) gripper control

    if ($z_{\text{ind}} < Tz_{\text{ind}}$) then

      Robot Gripper Close

    else

      Robot Gripper Open

end

Algorithm 3.1 describes the master–slave mapping algorithm logic and process steps of the proposed hybrid motion mapping algorithm GuLiM. Based on the predefined hand gesture rules in Sect. 3.2 of this chapter, the operator can freely adjust to a comfortable posture before the teleoperation starts. As shown in part (i) of the algorithm flow, first obtain the current motion state of the operator and the robot, including the operator's hand gestures, upper limb movements, and the current joint space data and end effector pose data of the robot. When a bending signal is detected in the operator's middle finger, calculate the relative pose $p_{\delta 1}$ between the robot and the operator, as shown in part (ii) of the algorithm flow. When the operator releases the middle finger, the motion of the robot arm stops, and the operator can freely adjust to a comfortable posture. When the middle finger is bent again, the robotic arm can move to transfer the human–robot action mapping and calculate the new human–robot relative posture. Continuous adjustment is applicable when the range of motion of the operator's own limbs has reached its limit, but the robot has not reached the target point, or the operator wishes to stop the mapping and rest during the operation. As shown in part (iii) of the algorithm flow, based on the proposed incremental posture mapping strategy, calculate the robot's target posture for the next movement based on the obtained relative posture. Then, enter part (iv) of the algorithm flow, use the *trac_ik* open-source inverse solution algorithm [2] to calculate the robot's target joint space, transmit it to the robot end controller for joint position control, until the robot reaches a suitable grasp position and grasp posture near the grasping target. At this point, the operator can bend the index finger to control the robot gripper for grasping, as shown in part (v) of the algorithm flow, where the opening and closing control of the gripper is an independent program thread in the control program, running parallel to the motion control of the robotic arm.

Figure 3.5 illustrates the intuitive process of the proposed human–robot action hybrid mapping method based on gesture data. According to the predefined gesture rules described above, the operator freely adjusts their posture in stage (i), followed by initiating pose mapping by bending the middle finger. In stage (ii), the relative pose is obtained and pose mapping teleoperation is performed based on the proposed incremental posture mapping strategy. In stage (iii), the operator's range of motion has reached its maximum limit, but the robot has not reached the target point, or the operator finds it inconvenient to operate in the current posture. The operator then releases the middle finger, no longer transmitting the motion to the robot, and freely adjusts to a suitable posture. Similar to stage (ii), in stage (iv), the operator bends the middle finger to continue the motion transmission and calculates a new relative pose at the beginning. This process continues until the robot reaches the target point with an appropriate grasping posture. In stage (v) and (vi), the operator bends the index finger to control the robot's servo gripper, where the arm control and gripper control are independent during the teleoperation process.

Building upon the coupled hand gesture incremental posture mapping strategy, the mapping command switching between gripper control and arm motion control during single-arm posture mapping teleoperation has been achieved. To further implement dual-arm posture mapping, the operator's left and right arm end position data $P_h = (x_h, y_h, z_h)$ is mapped to the position of the YuMi robot's left and right end effectors, denoted as$P_r = (x_r, y_r, z_r)$, as shown in Fig. 3.6. Similar to single-arm posture mapping, the initial coordinate systems $\{F_{\text{handR}}\}$ and $\{F_{\text{handL}}\}$ of the operator's right and left hands are not aligned with the coordinate systems $\{F_{\text{gripperR}}\}$ and $\{F_{\text{gripperL}}\}$ of the robot's right and left end effectors. To map the orientation of the operator's hand coordinate systems to the end effectors, new right and left hand coordinate systems $\{F_{\text{handRM}}\}$ and $\{F_{\text{handLM}}\}$ are created by rotating the original right hand coordinate systems $\{F_{\text{handR}}\}$ and left hand coordinate systems$\{F_{\text{handL}}\}$, respectively. These new coordinate systems have the same position as the initial coordinate systems but different orientations, aligning them with the orientations of the end effectors. This allows for direct implementation of incremental posture mapping

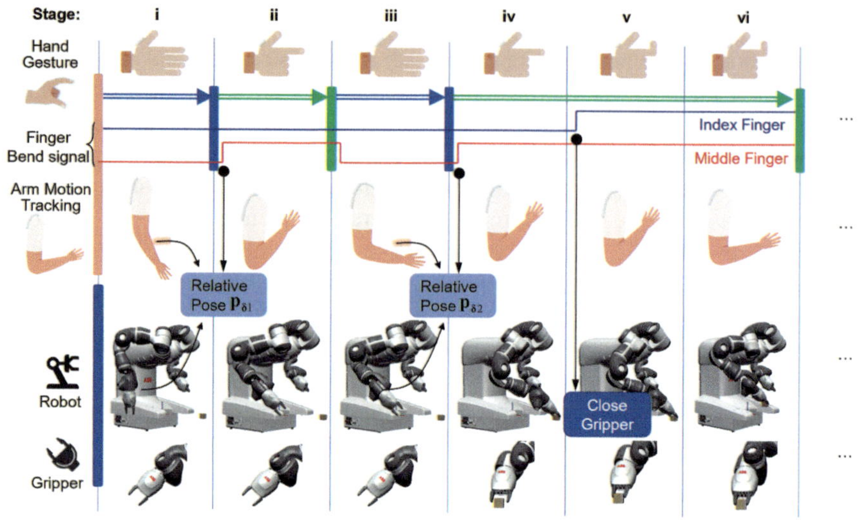

Fig. 3.5 The process of the GuLiM human–robot motion mapping technique

coordinate transformations during the mapping process. To facilitate flexible coordination between the left and right arms in posture mapping, a gesture-triggering rule for hand dominance swapping is designed for dual-arm posture mapping teleoperation. The bending state of the operator's right middle finger is used to enable or disable the motion control and teleoperation action transmission of the robot's left arm, and similarly, the bending state of the operator's left middle finger is used to enable or disable the motion control and teleoperation action transmission of the robot's right arm. Operators following this rule in dual-arm posture mapping teleoperation do not experience erroneous operations due to simultaneous operation of the arms and end effectors with the same hand dominance, improving the flexibility and reliability of the motion output control.

3.3 Comparison of Incremental Mapping and Direct Mapping Method

To validate the performance of the proposed GuLiM method, this section conducts comparative experiments. The Directly Mapping Method (DMM) is chosen as the comparative benchmark. The DMM mapping method directly maps the operator's hand posture to the robot's end effector without considering the relative coordinate transformation between the human and the robot during the mapping process. It is the most common and simple mapping method used in posture mapping teleoperation. The comparative experiment involves both an experienced operator familiar with the system and a novice operator who has never used the teleoperation system before. The

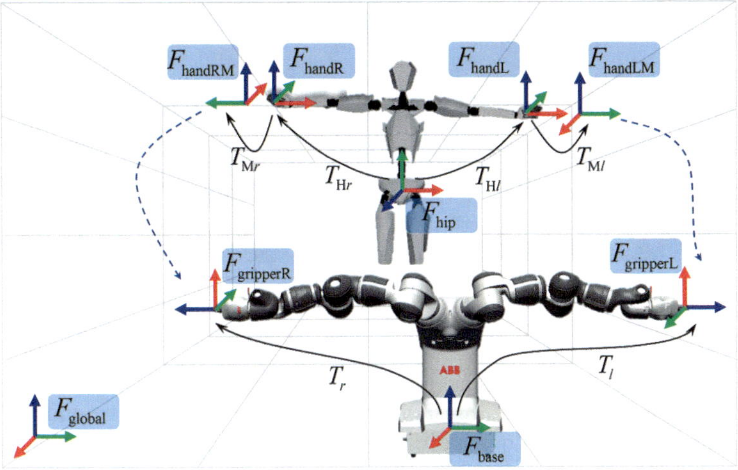

Fig. 3.6 Dual-arm human–robot coordinate transformation in Cartesian space

experimental task involves a common human task of grasping and placing blocks, requiring the operator to grasp a block from one location and place it at a specific position or orientation. To comprehensively evaluate the teleoperation control performance, this section proposes evaluation metrics based on task placement accuracy and time cost, and analyzes the experimental results.

3.3.1 Analysis of a Grasp Task Using Incremental Motion Mapping

A typical operation process for picking up and placing based on the proposed hybrid mapping method is shown in Fig. 3.7. The operator stands beside the robot and starts from a suitable operating posture, bending the middle finger to enable the human–robot motion mapping. Then, the operator moves and controls the end effector to approach the block. However, during this process, the position of the block in the operator's hand movement space (the position of the end effector) has exceeded the limit of the operator's range of motion, making it impossible to control the robot's end effector to reach the position of the block. The proposed hybrid mapping method can effectively solve this problem. The operator releases the middle finger at this time, and the human–robot motion mapping is no longer performed. Then, the operator retracts the arm and adjusts to a position and posture where movement can be continued. After that, bending the middle finger re-enables the human–robot motion mapping, until the end effector reaches the target position, and then bending the index finger controls the opening and closing of the gripper to pick up the target block. Similarly, during the placement process, the operator finds it difficult to adjust the arm movements of

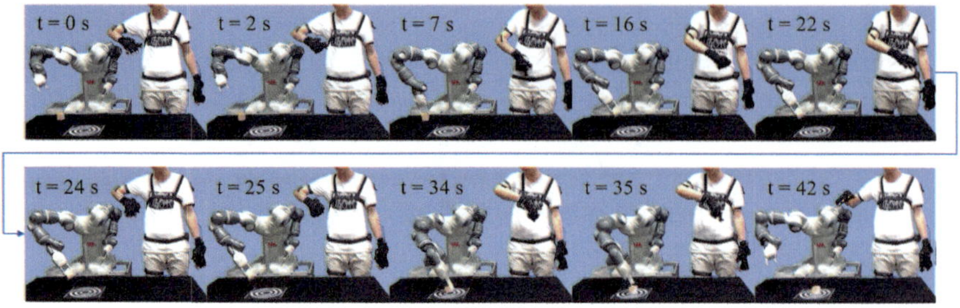

Fig. 3.7 Frame flowchart of grasping using GuLiM method, reprinted from Lyu et al. [1], Copyright (2025), with permission from IEEE

the robot again at t = 22 s. At this moment, the operator releases the middle finger, and the human–robot motion mapping ceases. The operator then retracts the arm at t = 24 s to a suitable position for continued operation. Subsequently, by bending the middle finger, the human–robot motion mapping is reactivated at t = 25 s, continuing until the end effector reaches the target position at t = 34 s, and the gripper is opened to release the block at t = 35 s.

Figure 3.8 shows the recorded motion signals of the operator and the robot during the experimental operation process corresponding to the GuLiM mapping teleoperation in Fig. 3.7. The bending signals of the index finger and middle finger of the operator are shown in the upper subgraph of Fig. 3.8. The bending signal of the index finger is used to control the end effector of the robot, while the bending signal of the middle finger is used to enable the transmission of motion commands between the human and the robot. These two finger bending gesture signals are independent of each other and triggered based on the previously mentioned bending state thresholds. In addition, the hand position data of the operator and the end effector position data of the robot are also recorded, as shown in the middle and lower subgraphs of Fig. 3.8. Before time t_1 in Fig. 3.8, as $z_{\text{mid}} < Tz_{\text{mid}}$, the robot's motion control was not enabled, and the robot remained stationary. During the time interval $[t_1, t_2]$, the operator bent the middle finger to $z_{\text{mid}} > Tz_{\text{mid}}$, enabling the transmission of motion commands and mapping the hand motion data to the robot end effector using the incremental posture mapping strategy. At time t_2, the operator's upper limb has reached the limit of the motion space, based on the hand gesture mixed incremental posture mapping strategy proposed in this section, the operator relaxed the middle finger at time t_2 to stop the robot's motion control and teleoperation mapping, during the time interval $[t_2, t_3]$, the robot remained stationary, and the operator could freely adjust the position of the hand without changing the robot's motion state, reaching a posture with convenient operation and redundant motion space. the operator completed the pose adjustment at time t3, and then bent the middle finger again to $z_{\text{mid}} > Tz_{\text{mid}}$, re-enabling the teleoperation control of the robot, transmitting the operator's hand relative motion state. This process continues until the manipulator reaches the appropriate grasping target point under the operator's teleoperation control. At this point, the operator

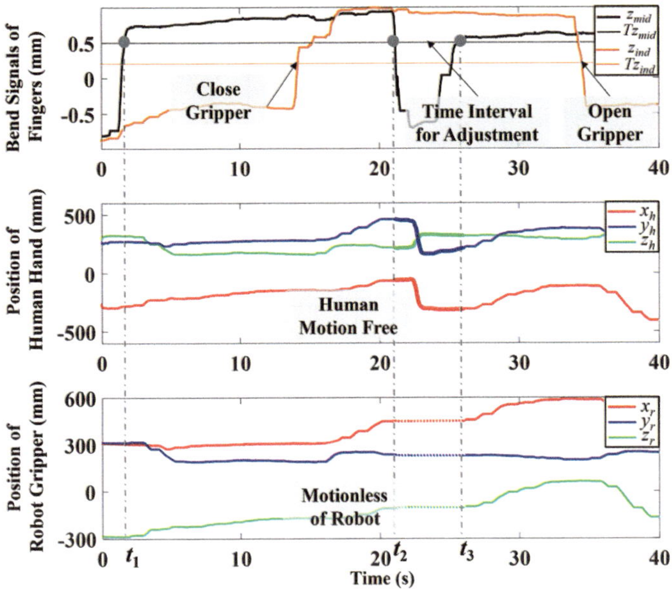

Fig. 3.8 The motion data curves in the GuLiM process, reprinted from Lyu et al. [1], Copyright (2025), with permission from IEEE

bends the thumb to $z_{\text{ind}} < Tz_{\text{ind}}$ opening the end effector of the manipulator and performing the grasping task.

3.3.2 *Definition of the Evaluation Metrics and Experimental Setup*

In this study, the traditional DMM (Directly Mapping Method) was selected as the experimental benchmark to compare the performance of the proposed hybrid mapping method. Two sets of experiments for picking up and placing blocks were conducted to validate the practical performance of the hybrid mapping method. During the experiments, the operators were required to pick up blocks from region A, as shown in Fig. 3.9, and place them into one of the target regions, namely regions B, C, and D.

To evaluate the grasping performance of the proposed method in this chapter, a grasping performance evaluation metric is proposed in two sets of experiments. It consists of two main parts: the precision of placement ($S_{mp}, m = 1, 2$) and the operation time to complete the task, where m denotes the mth group of experiments. This section uses S_{1p} to represent the placement precision performance score of the first group of experiments and S_{2p} to represent the placement precision performance score of the second group of experiments.

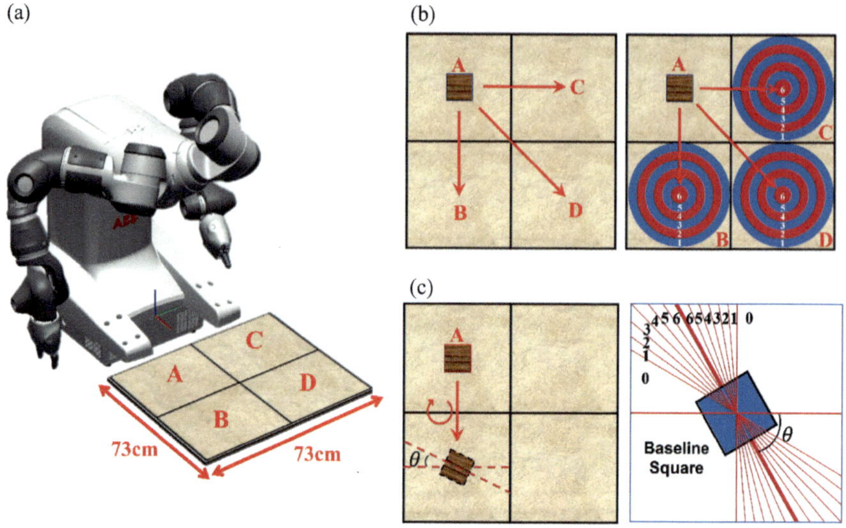

Fig. 3.9 Experimental setup and illustration of accuracy evaluation, reprinted from Lyu et al. [1], Copyright (2025), with permission from IEEE. **a** Pick-and-place experiment setup. **b** Position accuracy evaluation setup. **c** Orientation setup

In the first group of experiments, as shown in Fig. 3.9b, each target area was divided into a pair of concentric rings as scoring rings. Each ring corresponds to a level of evaluation score, where a larger diameter corresponds to a further outward deviation of the placement point, resulting in a lower score, the score range is set from 1 to 6. For the target areas B, C, and D, five pick-and-place experiments were conducted for each. In the second group, as shown in Fig. 3.9c, the operator placed the block into the target region with a specified angle $\theta(\theta = [30°, 60°, 90°])$. Six sector areas were divided based on the angle deviation with scoring S_{2p}^{dev} ranged from 1 to 6. The scoring areas were symmetrically distributed on the left and right sides of the datum line, and the area with a larger deviation would lead to a lower score. Besides, the position precision of placement was also considered with four scoring levels ranging from 1 to 4. This part of the score was based on the ratio ($\alpha = A_{overlap} / A_{square}$) of the overlapping area $A_{overlap}$ between the block and the baseline square, to the area of the baseline square A_{square}. The final score of the placement precision of the second group S_{2p} was calculated with the scoring rules depicted below in (3.12).

$$S_{2p} = \begin{cases} S_{2p}^{dev} \times 0 & 0 < \alpha \le \frac{1}{3} \\ S_{2p}^{dev} \times 1 & \frac{1}{3} < \alpha \le \frac{1}{2} \\ S_{2p}^{dev} \times 2 & \frac{1}{2} < \alpha \le \frac{2}{3} \\ S_{2p}^{dev} \times 3 & \frac{2}{3} < \alpha \le 1 \end{cases} \tag{3.12}$$

The final score is calculated by multiplying the two parts of the score as described above. Unlike the first set of experiments, this set mainly focuses on the accuracy of the placement direction. Therefore, only one target area (area B) was selected, but three different placement directions ($\theta = [30°, 60°, 90°]$) were chosen. For each angle, five pick-and-place trials were conducted.

During the experiments, as described above, the placement accuracy scores of the two groups of experiments were recorded. The evaluation of operational efficiency in completing teleoperation tasks directly recorded the time consumed for each task (unit: seconds) and conducted analysis and comparison. To assess the user-friendliness of the proposed GuLiM intuitive mapping method for novice operators, each group of experiments (including the placement accuracy evaluation experiment and the placement posture accuracy evaluation experiment) was completed by the aforementioned experienced operator and novice operator for subsequent comparison and analysis.

3.3.3 Placement Accuracy Evaluation for Grasp Task

The results of the first group of experiments used to evaluate the placement accuracy are shown in Fig. 3.10a, where P1 and P2 represent two operators involved in the experiments. P1 is well-trained and familiar with the operating system, while P2 is a novice who encounters this teleoperation system for the first time. From the experimental results, it can be seen that the proposed GuLiM mixed intuitive mapping method outperforms the traditional DMM direct mapping method in terms of placement accuracy scores. Additionally, the GuLiM method has a smaller standard deviation in scores compared to the DMM method, indicating its robustness. Specifically, in the grabbing and placing tasks from area A to D, GuLiM significantly outperforms DMM in placement accuracy scores, as the grabbing and placing operations between area A and D are the largest distance tasks among the three. When using DMM, the larger operating distance may cause the operator's hand to reach the limit of their own body's motion space, making it difficult for the operator to place the grasped object at the target position. On the other hand, in the simpler tasks of the first group of experiments, DMM is more efficient than the GuLiM method. This is because the DMM method does not require an adjustment process compared to the GuLiM method, saving time during the adjustment process, which is also reasonable. In the second group of experiments, according to the task design mentioned earlier, the object was grabbed and placed at a specified angle $\theta(\theta = [30°, 60°, 90°])$, and the S_{2p} score was calculated according to the scoring rules. As shown in Fig. 3.10b, in all experiments evaluating placement posture accuracy, the scores using the GuLiM intuitive mapping method were superior to those using DMM.

Figure 3.11 summarizes the improvement percentage of precision and time cost of GuLiM compared with DMM. According to the statistical results, as shown in the left diagram of Fig. 3.11, the GuLiM surpassed DMM by 46.77% in terms of placement precision averagely, whereas it took 19.60% more time on average to accomplish the tasks. For novice operator P2, the GuLiM method has a larger improvement in placement precision than the DMM method compared with the experienced operator.

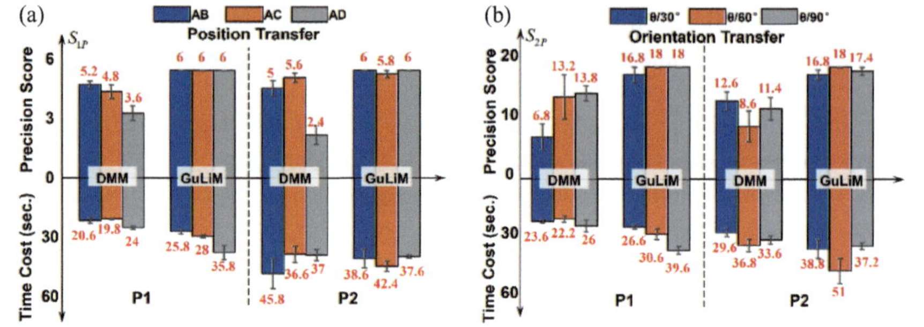

Fig. 3.10 Placement accuracy and time consumption for different operators. **a** Results of position accuracy evaluation. **b** Results of orientation accuracy evaluation

As shown in the right diagram of Fig. 3.11, in the evaluation of pose placement accuracy experiment, the S_{2p} score of the GuLiM surpassed that of the DMM by 69.27% but took 30.54% more time on average for orientation transfer.

From the above experimental results, it is well-reasoned to say that the proposed GuLiM method is capable of accomplishing tasks with more complicated operations. Besides, the GuLiM method takes a main advantage on the placement precision. However, the average time of the GuLiM method is slightly increased than the DMM method. Reviewing the mixed pose mapping teleoperation process with hand gestures and upper limb motions shown in Figs. 3.7 and 3.8, the reason why using the GuLiM method leads to an increase in time cost is that this method requires the operator to make necessary adjustments in hand gestures and body posture during the teleoperation grasping and placing tasks, which adds additional time cost. An average of two adjustments are required in the position transfer assessment tasks (AB, AC, and AD), while an average of three adjustments are required in the orientation transfer assessment tasks ($\theta = [30°, 60°, 90°]$). Therefore, the average time of the GuLiM method is slightly increased due to the adjustment procedure. However, the GuLiM method does not require complicated setup and calibration procedures

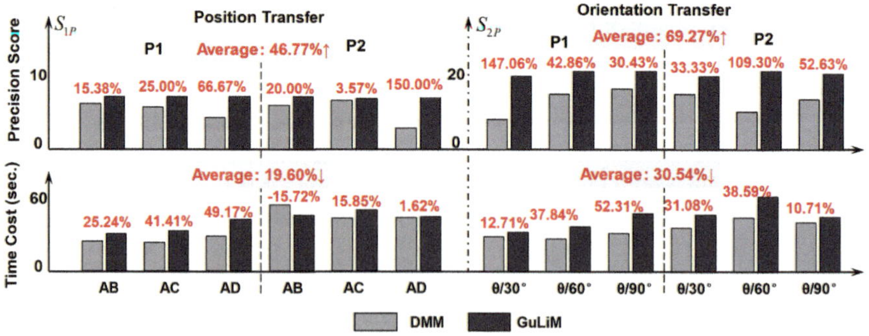

Fig. 3.11 Comparison of the GuLiM and DMM in task accuracy and efficiency

before an operational task, which saves the pre-training time of the operators and improves its usability and generalizability in practical application scenarios.

3.4 Summary

The intuitiveness and convenience of teleoperation systems are key in the practical application of robot teleoperation systems. Designing user-friendly and efficient teleoperation strategies for users unfamiliar with robot operations has always been a significant challenge in the promotion of teleoperation technology. Building upon the implementation of single-arm posture mapping teleoperation by capturing operator upper limb movements, this chapter proposes a GuLiM dual-arm human–robot posture mapping method based on coupling hand gestures and upper limb movements. This method effectively leverages the subjective intervention ability of the master operator during the teleoperation process, improves the intuitiveness of dual-arm robot posture mapping teleoperation, and significantly reduces the training and learning costs for the master operator. Additionally, an incremental posture mapping strategy is proposed, which not only make full use of the robot arm's workspace but also simplifies the pre-calibration process in human–robot posture mapping teleoperation, enhancing the convenience of the posture mapping process. Furthermore, a comparative experiment is conducted to evaluate the performance differences between the GuLiM mapping method proposed in this chapter and the traditional direct mapping method DMM in grasping and placing actions. The main conclusions of this chapter are as follows:

(1) A proposed incremental posture mapping strategy is introduced, which applies real-time calculation of the relative pose transformation between the robot and the operator to the kinematic solution of the target pose. This strategy achieves the mapping and coordinate transformation of human–robot equivalent postures in Cartesian space, eliminating the cumbersome initialization and calibration process in traditional teleoperation. It also allows for flexible adjustment of the operator's own limb posture during teleoperation, addressing the limitation of the operator's motion range in the master–slave heterogeneous human–robot posture mapping operation. This strategy improves the convenience and intuitiveness of human–robot posture mapping teleoperation.
(2) Building on the design of the incremental posture mapping strategy, this chapter proposes the GuLiM posture mapping method based on a mix of operator hand gestures and limb movements. This method allows for more flexible intervention in the motion transmission process by incorporating predefined hand gestures while transferring the operator's upper limb movements to the robot. This enables coordinated operation control between the robotic arm and the end effector during posture mapping teleoperation. By designing gesture-triggering rules that swap hand dominance, the method achieves dual-arm posture mapping

teleoperation, further ensuring the reliability and flexibility of the output of the dual-arm teleoperation control.

(3) The experiment designed for grasping and placing tasks quantitatively evaluates the performance of GuLiM posture mapping method compared to the traditional DMM direct mapping method, focusing on placement accuracy and operational efficiency. The results shows that the GuLiM method improved placement accuracy by 46.77% in position placement and 69.27% in posture placement compared to the DMM method. However, the GuLiM method takes 19.60% longer for position placement evaluation and 30.54% longer for posture placement evaluation compared to the DMM method. The GuLiM method demonstrates significant advantages in placement accuracy, although it slightly increases the average time for operators to complete tasks. Furthermore, comparing the performance of proficient operators and novices, the results shows that the proposed GuLiM mapping method can help novice operators improve position and posture placement accuracy during teleoperation.

Reference

1. Denavit J, Hartenberg RS (1955) A kinematic notation for lower-pair mechanisms based on matrices. J Appl Mech
2. Beeson P, Ames B (2015) TRAC-IK: an open-source library for improved solving of generic inverse kinematics. In: Proceedings of the international conference on humanoid robots, Seoul, Korea, pp 928–935

Chapter 4
Whole-Body Human-Motion Based Robot Teleoperation

Abstract This chapter addresses the challenge of teleoperation for robots designed for highly dynamic tasks, such as homecare or medical assistance, which often include not only robotic arms but also mobile chassis and redundant body motion mechanisms (e.g., humanoid torsos and wheeled chassis). The primary challenge in pose mapping teleoperation is extending control beyond the arm to include the robot's mobile chassis and other body motions. To meet these needs, the chapter proposes a dynamic gesture trajectory recognition method that captures the operator's body movements, allowing control of the robot's mobile chassis and torso motions. The corresponding gesture mapping strategy is designed for the robot's mobile control to verify the method through functional tests on an integrated nursing assistant robot platform. Additionally, a humanoid dual-arm robot with redundant torso degrees of freedom is used, with the operator's torso and lower limb movements captured to enable whole-body cooperative teleoperation control. The system is then validated in two key healthcare scenarios: remote homecare (Healthcare 4.0) and telemedicine assistance. This approach demonstrates how the proposed teleoperation system can enhance robot control flexibility and effectiveness in real-world healthcare applications.

Keywords Whole-body motion mapping · Gesture recognition · Mobile manipulation · Robot torso · Redundant DoFs

4.1 Introduction

In addition to the robotic arm actuator, which is responsible for the execution of the operation task, robots that perform such highly dynamic and strong environmental interaction tasks like homecare or medical assistance generally have a mobile chassis and other redundant degrees of freedom proprioceptive motion mechanisms to control the movement of the robot and the adjustment of the robotic arm's workspace, and most of them are designed with humanoid torso mechanisms and wheeled mobile chassis [1, 2]. When the pose mapping control of the robot arm is completed, in

H. Lyu et al., *Human Motion Awareness and Robot Teleoperation*, Advanced Topics in Science and Technology in China 4, https://doi.org/10.1007/978-981-96-6545-7_4

the process of landing the pose mapping teleoperation application, the key problem is to utilize the operator's body movements to achieve other robot proprioceptive motions except the robot arm actuator. In this chapter, a dynamic gesture trajectory recognition method of the operator's terminal was proposed to meet the teleoperation control requirements of the mobile chassis of such a nursing assistant robot and the redundant body motion mechanism like the human torso. The corresponding gesture mapping strategy was designed for the robot's mobile control, and functional verification were carried out on an integrated design nursing assistant robot platform. In addition, a humanoid dual-arm robot with redundant torso degrees of freedom was integrated, and the corresponding torso pose mapping strategy was designed by capturing the operator's torso motion and lower limb motion to realize whole-body cooperative teleoperation control. Finally, for the two typical application scenarios of Healthcare 4.0 remote homecare and telemedicine assistance, the developed posture mapping teleoperation system was verified and applied based on the self-built nursing assistance robot platform.

4.2 Human-Motion Based Teleoperation of Robot Mobile Chassis

Human gesture is an important way for human to express their intentions to the outside world. Therefore, gesture recognition technology provides a feasible solution for convenient human–computer interaction [3]. Most current gesture recognition systems rely on complex external equipment, such as indoor optical 3D motion capture systems [5] and other fixed visual capture solutions [4]. In this section, a lightweight wearable wrist camera was designed for the convenience requirements of human–computer interaction. By analyzing the real-time image background data of the wrist camera, the motion data of the background feature points in the process of the operator's gesture is extracted, and the motion trajectory of the operator is calculated backward. The block diagram of the human–computer interaction process based on dynamic gesture trajectory recognition is shown in Fig. 4.1. In this section, a series of gesture instructions were designed and developed for the mobile interaction requirements of nursing assistant robots, and the classification and recognition of corresponding gesture instructions are realized based on the K-nearest neighbor algorithm. At the same time, a trajectory recognition experiment was designed to verify the performance of the gesture recognition system, and the teleoperation function of the mobile nursing robot's motion control was also tested.

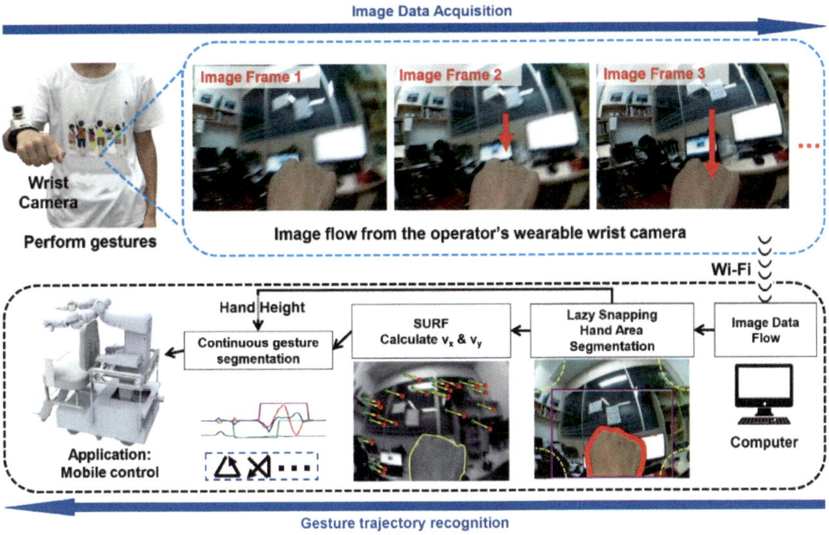

Fig. 4.1 Block diagram of the HRI system based dynamic gesture recognition

4.2.1 Dynamic Gesture Trajectory Recognition

4.2.1.1 Hardware Structure Design of Wearable Wrist Camera

In order to achieve the recognition of the continuous dynamic gesture trajectory of the operator, a wearable wrist camera was designed in this section, and the hardware structure is shown in Fig. 4.2. Based on 3D design software, we designed lightweight and foldable rotating shaft structure for the device, which can be easily stored and carried around, as shown in Fig. 4.2a. The base, housing, and camera cover of the wearable wrist camera are 3D printed in nylon material (housing thickness is 1 mm), and the overall weight of the wearable wrist camera is low (the final prototype device weighs 114 g). The foldable structure of the camera makes the overall structure compact, while the base and elastic wearable wristband are bonded by hot melt glue. In addition, the pull-out structure design between the shell and the base makes it convenient for users to disassemble and carry. The overall dimensions of the designed wearable wrist camera are shown in Fig. 4.2a. The camera module is fixed in the camera housing, which is hinged between the main housing by a rotating shaft and can be flipped from 0° to 150°. In normal gesture recognition, the reversal Angle is set to 80°, which can be folded and stored in non-working state.

The functional hardware composition and prototype of the above wearable wrist camera are shown in Fig. 4.2b. The camera sensor selected is a USB camera module compatible with the Raspberry PI, which is a CMOS 175° wide-angle camera with a resolution of 5 megapixels (2952 × 1944 pixels). Raspberry PI Zero W is selected as the controller, which integrates a single-core central processing unit (CPU) at 1

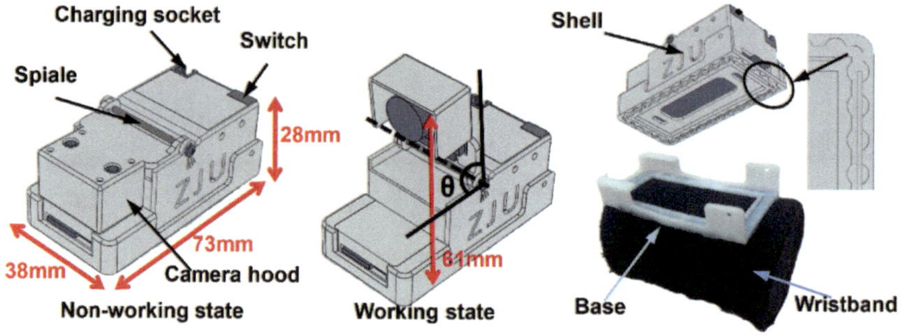

(a)

(b)

Fig. 4.2 Structure and hardware composition of the wearable wrist camera, reprinted from Ref. [6], CC BY 4.0. **a** The hardware composition of the wearable wrist camera. **b** Design of wearable wrist camera prototype

Ghz and 512 MB of RAM, and supports 802.11 b/g/n wireless LAN connection. The controller module is small (65 mm × 30 mm × 5 mm) and has a higher clock frequency compared with other microcontroller modules such as Arduino, making it suitable for high-speed image acquisition and transmission. The USB camera module and the Raspberry Pi controller are connected by a flexible flat cable. According to the power requirements and the size constraints of the integrated design, two rechargeable Li-ion batteries with a rated voltage of 3.7 V are selected for parallel output, with a total capacity of 2000 mAh. Meanwhile, in order to satisfy the power requirements of the Raspberry Pi and the camera module, a boost module is used to obtain a voltage output of 5 V.

The acquisition of the image-video streaming data is achieved employing a Python program script running on a Raspberry Pi Raspbian system (a Debian GNU/Linux-based Raspberry Pi hardware development system). Firstly, a TCP connection to the host computer is established. After successfully connected, the Raspberry PI collects video image data from the camera module and transmits it to the host computer in real-time. In order to meet the processing requirements of the algorithm, the transmitted image quality is 320 × 240 pixel resolution, and the transmission speed is set to

12 FPS. Secondly, in terms of the power consumption of the whole machine, the operating current is 0.21 A in the video transmission state and 0.11 A in the boot state without data transmission. Finally, according to the battery capacity and the actual use test, the device can work continuously for more than four hours to meet the use requirements in common human–computer interaction scenarios.

4.2.1.2 Process Design of Dynamic Trajectory Gesture Recognition Algorithm

After the original image data is obtained by the host computer, the data is first decoded and converted, and then processed by the subsequent recognition algorithm. When the operator makes gestures with the wrist camera, the user's hand is in the middle of the picture will be captured by the camera. Compared with the change range of the overall picture, the change characteristics of the picture are more obvious in the background picture around the user's hand. Therefore, the feature change of the background in the picture can be selected to infer the operator's hand movement, so as to indirectly reflect the actual movement trajectory of the operator's hand. In addition, in order to distinguish valid gesture trajectories from other invalid gestures during the continuous motion of the operator, this paper designed a method to segment gestures by detecting the swing of the operator's hand. The operator bends his wrist and swings it toward the bottom of the camera screen before making a valid gesture. Then, the algorithm detects the height change of the hand region in the image to segment the gesture interval. Due to the difference in hand size between operators, the operator needs to ensure that the hand disappears from the screen when the wrist is bent. After obtaining the effective gesture data, the gesture recognition results are obtained by classifying the gestures based on DTW similarity matching algorithm.

As shown in Fig. 4.3, the algorithm flow scheme showed how to realize continuous dynamic gesture trajectory recognition, which is mainly divided into the following four steps; (1) hand region segmentation to obtain the height features of the hand; (2) background speed calculation, which can be obtained by matching background feature points between successive frames; (3) continuous gesture segmentation, achieved by detecting changes in the height of the hand area caused by bending the wrist; (4) data normalization and cross-validation classification, which can test the recognition accuracy to evaluate the performance of the system.

(1) **Hand region segmentation**:

As mentioned above, in the process of continuous gesture recognition, the motion of the operator's wrist swing is recognized through the height change of the hand area in the image, thus triggering one gesture track recognition. In addition, when calculating the speed of the background feature, the foreground area of the hand should be removed, leaving only the background part for feature point matching. Therefore, the recognition and segmentation of the hand region is the key to the algorithm framework.

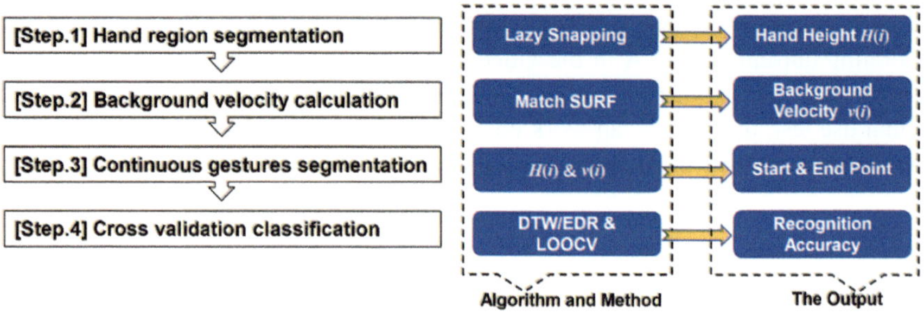

Fig. 4.3 Process of dynamic gesture trajectory recognition algorithm, reprinted from Ref. [6], CC BY 4.0

In order to improve the image processing efficiency and recognition speed of the algorithm, the original image is first downsampled from 240 × 320 pixels to 72 × 96 pixels. The wearable wrist camera designed in this book uses a wide-angle camera module, so the original image acquired by the camera has wide-angle distortion, as shown in Fig. 4.4a. The wide-angle distortion is generally corrected by the corresponding correction algorithm [7]. However, in this study, the background screen outside the hand area in the image is important raw data for speed calculation. To avoid feature point correction errors, here does not apply the commonly used wide-angle distortion correction algorithm but uses the pixel clipping method to remove the four corners with serious distortion in the image, and cuts the image pixels to 54 × 73 pixels, as shown in Fig. 4.4b. To further improve the performance of the algorithm for subsequent processing, the RGB image is converted to $L^*a^*b^*$ color space image as shown in Fig. 4.4c. Then, Simple Linear Iterative Clustering (SLIC) algorithm [8] is used to generate compact and nearly uniform super-pixel images, as shown in Fig. 4.4d, to further reduce the computational cost of hand region segmentation.

After the pre-processing of the image in Fig. 4.4a–d, this book used the Lazy Snapping algorithm [9] to separate the foreground and background of the hand area. Lazy Snapping is an interactive image segmentation algorithm that segments the foreground and background of an image based on specified seed pixels. In the footage captured by a wearable wrist camera, the middle hand region foreground is clearly distinguished from the rest of the ambient background region, and hence it is suitable to segment the video image sequence by giving initial foreground and background seed pixels. Here, the initial seed pixels are given in Fig. 4.4e, the green area is the initial foreground seed of 8 × 13 pixels $S_{fore}(0)$, and the blue area is the initial background seed of 42 × 47 pixels $S_{back}(0)$. Further, in the successive image processing process, by giving the foreground seed $S_{fore}(i-1)$ and the background seed $S_{back}(i-1)$ of the previous frame, the hand region segmentation result $R_{hand}(i)$ of the current i frame can be obtained, that is, the white area and its internal area in Fig. 4.4f. The values of $S_{fore}(i-1)$ and $S_{back}(i-1)$ are jointly determined by the values of $R_{hand}(i-1)$, $S_{fore}(0)$ and $S_{back}(0)$: the foreground seed pixel $S_{fore}(i)$ is

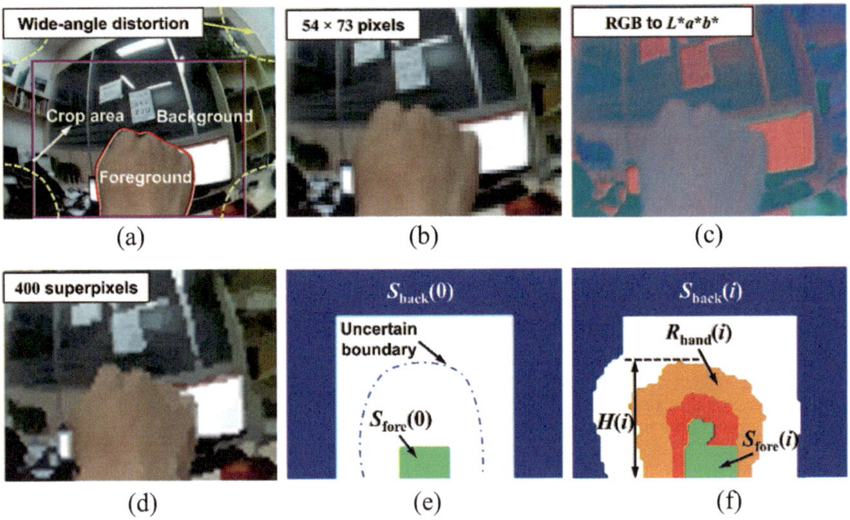

Fig. 4.4 Algorithmic processing flow of hand region segmentation, reprinted from Ref. [6], CC BY 4.0. **a** Original image and element composition. **b** Pixel cropped image. **c** $L * a * b*$ color space image. **d** Superpixel image. **e** Initial seed pixels. **f** Hand area segmentation

obtained by merging the eroded regions of the $S_{fore}(0)$ and $R_{hand}(i)$ pixels and is used for the gesture segmentation in the next frame, that is, the green area in Fig. 4.4f; the background seed pixel $S_{back}(i)$ is obtained and used in the next gesture segmentation by combining the corrupted regions of and $S_{back}(0)R_{hand}(i)$ pixels, namely the blue area in Fig. 4.4f. By analogy, the hand region segmentation in the whole dynamic trajectory image video stream can be accomplished through iteration. After obtaining the segmentation result of the hand region in each frame, one can be used for the background speed calculation, and the other can be used to calculate the hand height $H(i)$ in the current frame through the highest pixel point of the hand area, which can be used for judging the wrist swing action and the subsequent successive gesture segmentation processing.

As shown in Algorithm 4.1, in order to further reduce the impact of segmentation errors, $R_{hand}(i)$ was eroded inward by $Rows/8$ pixels to remove the noise, where $Rows = 54$ is the height of the pixel after the cropping process and the result of the erosion is $Re_{hand}(i)$ as shown in the red area of Fig. 4.4f. If the area of $Re_{hand}(i)$ is 0, the height $H(i)$ is equal to the height of $R_{hand}(i)$; if the area of $Re_{hand}(i)$ is not 0, the height is equal to the height of $Re_{hand}(i)$ plus the distance of the erosion $Rows/8$. Finally, the calculated height was passed through a mean filter to obtain the final output height $H(i)$.

Algorithm 4.1 : Segmenting the hand region R_{hand} and calculating its height H

Input: Continuous Image Video Stream, $S_{fore}(0)$, $S_{back}(0)$

Output: $R_{hand}(i)$, $H(i)$

for $i = 1$ to (Number of images in the video sequence)

    (I) Image pre-processing, pixel reduction

    $\mathrm{Img}(i) \leftarrow \mathrm{ReducePixels}(\mathrm{Img}_{source}(i))$

    (II) Hand region image segmentation

    $Rows \leftarrow \mathrm{NumerofRows}(\mathrm{Img}(i))$

    $R_{hand}(i) \leftarrow \mathrm{LazySnapping}(\mathrm{Img}(i), S_{fore}(i-1), S_{bcak}(i-1))$

    $S_{back}(i) \leftarrow \mathrm{AND}(S_{back}(0), \mathrm{NOT}(\mathrm{Dilate}(R_{hand}(i), Rows/5)))$

    $S_{fore}(i) \leftarrow \mathrm{OR}(S_{fore}(0), \mathrm{Erode}(R_{hand}(i), Rows/4))$

    (III) Hand height calculation in the image

    $H_0(i) \leftarrow \mathrm{CalcuHeight}(\mathrm{Img}(i))$

    $H(i) \leftarrow \mathrm{Mean}(H_0(i-2), H_0(i-1), H_0(i))$

end

(2) **Calculate the background speed**:

In this book, the operator's gesture trajectory is calculated with the background speed characteristics of the wearable wrist camera. After completing the above hand region segmentation steps the problem is to obtain the background speed in the image. The calculation of background speed depends on the selection of reference points. In this book, the Speeded Up Robust Features (SURF) algorithm was used to select background reference points [10], and the background speed is calculated by the average displacement of SURF feature points matched between two adjacent frames of images. The algorithm can detect the local extremum points of the second derivative of the image at different scales from the image, and then describe the characteristics of these points with gradient histogram. Such feature points have excellent properties such as robustness, scale invariance, and rotation invariance.

The specific algorithm process is as follows: First, SURF feature points of two adjacent frames of images are extracted, and remove feature points in the foreground hand region according to the segmentation results of the hand region, while only feature points in the environmental background are retained. As shown in Fig. 4.5a and b, the green + symbol and the red dot symbol are the feature points extracted in the two neighboring image frames, respectively. After that, the Euclidean distance between the feature points of the current frame and the feature points of the previous frame are calculated refer to Loweet et al.'s method in the literature [11]. The feature points in the two adjacent frames are matched according to the principle of minimum distance (for a feature point, the feature point with the smallest distance in the other frame is taken as its matching feature point).

After obtaining the matching feature points between two adjacent frames, the velocity vector can be calculated based on the pixel displacement of the feature points, and the velocity vector of each feature point is shown by the yellow arrow annotation in Fig. 4.5c. Further, the velocity components V_x and V_y of each feature point in the horizontal and vertical directions of the picture are calculated. Some feature points located at the edge of the image may have matching errors, resulting

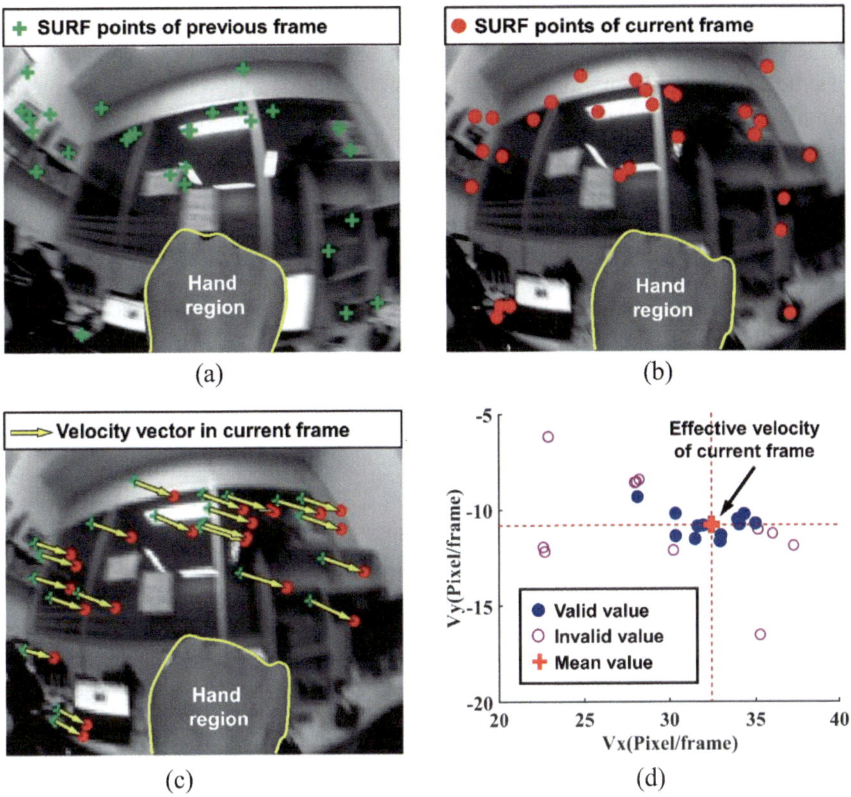

Fig. 4.5 Calculation of the background velocity of the key points, reprinted from Ref. [6], CC BY 4.0. **a** Keypoints in the previous frame. **b** Keypoints in the current frame. **c** Keypoints matching and velocity vector calculation. **d** Calculation of the effective background speed for the current frame

in excessively large or small velocity values. In this book, the feature points with the largest 20% and smallest 20% of the velocity values of V_x and V_y are regarded as invalid values, as shown in the pink circle symbols in Fig. 4.5d, and the remaining other valid values are shown by the blue circle symbols in Fig. 4.5d. Further, the average of the velocity values of effective feature points of the current frame is calculated as the background speed of the current frame, as shown in Fig. 4.5d.

(3) **Segmentation of continuous gesture trajectory**:

After obtaining the background velocity by the above method, the next problem to be solved is how to segment the valid gestures from the continuous dynamic gesture trajectories. In order to distinguish effective and ineffective gestures in continuous motion, the following rules were defined in this book to identify the beginning and end of effective gestures: The user should bend and swing the wrist downward before starting the gesture, and maintain a static state for more than half a second after the completion of the gesture. In the process of algorithm debugging, totally 1000 gesture trajectories were collected from five subjects, and the 1000 gesture trajectories were saved as a dataset for the test of segmentation effective gestures. The specific gesture segmentation process is implemented as follows: Due to the inconsistent movement habits of different operators, it is impossible to remain completely stationary when stationary movements are required. Therefore, in this section, the maximum stationary speed threshold A was set to make a judgment. If the average value of the background velocity of 6 consecutive image frames (0.5 s) is less than V_t, the operator is considered to be stationary at the moment of the current last frame. In order to determine the optimal value of the maximum stationary velocity threshold V_t for continuous gesture segmentation, the segmentation accuracy test was performed on the collected 1000 gesture dataset using different values of V_t. The experimental results are shown in Table 4.1. The results show that the optimal segmentation accuracy can be obtained when $V_t = 3$(pixels/frame).

In order to explain the segmentation process of continuous gesture trajectory more intuitively, the algorithm flow chart of triangular trajectory segmentation is interpreted as an example here, as shown in Fig. 4.6. Based on the segmentation results of the previous hand region segmentation steps, the change of the real-time hand region height $H(i)$ which represents the amplitude of wrist swing can be obtained. Meanwhile, the real-time velocity change of the operator's gesture trajectory, i.e., the velocity components in two directions V_x and V_y, can be obtained through SURF feature point matching. According to the definition of the aforementioned continuous gesture segmentation rule, when the operator makes a wrist swing action, the operator's hand height $H(i)$ in the screen will have continuous falling and rising edges, as shown in S1 stage of Fig. 4.6. Accordingly, it is assumed that the operator will start an effective gesture movement in the following time interval 1 (indicated by the green interval). In interval 1, if it cannot be recognized as a stationary state, it is considered that the operator's effective gesture starts, as shown in interval 2 (pink interval marking) in Fig. 4.6. The operator's stationary state is allowed between the operator's wrist swing after the S1 phase and the beginning of the S3 phase of the effective gesture trajectory, as shown in phase S2 in Fig. 4.6. In this book, the time window width of the effective gesture trajectory monitoring is set to 2 s. If the stationary interval S2 after the operator swings his wrist is longer than the time window width of the effective gesture trajectory monitoring, the recognition of the trajectory gesture will not be started and the monitoring of the wrist swing will be restarted, that is, the gesture two seconds after the wrist swing finished will be regarded as an invalid gesture track. The S3, S4, and S5 phases in the figure correspond to the three linear trajectories of the triangular gesture trajectory, respectively. In general, after the end of the valid gesture trajectory, the continuous stationary state lasting for one second and more than one second in the S6 stage will be judged

Table 4.1 Segmentation accuracies under different thresholds V_t

V_t (pixels/frame)	1	2	3	4	5
Segmentation accuracy (%)	83.1	97.1	99.2	98.3	97.8

as the end of the gesture recognition. After the start of the valid gesture trajectory, the stationary state lasting for less than one second will not be judged as the end of the valid gesture. So far the effective gesture trajectory segmentation in a continuous gesture trajectory has been completed, and an independent effective gesture trajectory can be obtained for subsequent gesture trajectory recognition.

(4) **Gesture track recognition**:

After obtaining the velocity time series data of a single valid gesture trajectory, the data needs to be classified and recognized to facilitate the interface design for subsequent human–computer interaction. Since the actions of different operators have different durations and moving speeds, the background velocity component sequence data $[V_{gx}(i)]$ and $[V_{gy}(i)]$ of the X-direction and Y-direction of the gesture trajectory g were normalized by linear interpolation and resampling before classifying the gesture trajectories, so as to reduce the influence of the differences between different experimenters. Here, the number of normalized sampling points was set to 30, and the two sequence data after normalization were obtained as $[F_{gx}(i)]$ and $[F_{gy}(i)]$. The normalized background velocity sequence data in X and Y directions were combined as $\boldsymbol{F_g} = \begin{bmatrix} F_{gx}(i) \\ F_{gy}(i) \end{bmatrix}$. The dynamic time planning algorithm (DTW) [12] was used to calculate the difference between it and all predefined gesture template trajectory velocity vectors, and the gesture category with the smallest difference (the smallest DTW distance) was selected as the final recognition result. The recognition formula for gesture trajectory category G is as follows:

$$G = \underset{G}{\operatorname{argmin}}(\mathrm{DTW}(\boldsymbol{F_g}, \boldsymbol{F_{TG}})) \tag{4.1}$$

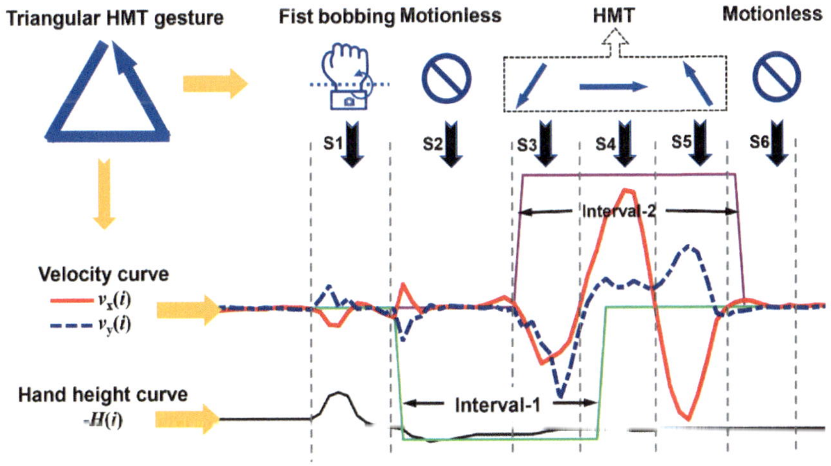

Fig. 4.6 Schematic diagram of the continuous gesture trajectory segmentation, reprinted from Ref. [6], CC BY 4.0

where $\boldsymbol{F}_g$ is the velocity sequence data of the input gesture g and $\boldsymbol{F}_{TG}$ is the velocity sequence data of the template gesture of category G.

4.2.2 *Hand Gesture Recognition for Robot Mobile Chassis Control*

According to the motion characteristics of the omnidirectional mobile robot in two-dimensional space, six motion modes of the omnidirectional mobile robot were firstly defined: forward motion, backward motion, left translation motion, right translation motion, clockwise rotation, and counterclockwise rotation. The above six basic motion modes can accomplish the flexible omnidirectional motion of the mobile robot [6].

In order to interact with the mobile robot using the recognition results of the dynamic trajectory of the gestures, this book defined 10 corresponding dynamic trajectory gestures for the motion control of the mobile robot, as shown in Fig. 4.7. The first six of them (gestures ①–⑥) were used for the six basic motion modes mentioned above. In addition, gestures ⑦ and ⑧ were defined for speeding up and slowing down, respectively. Gesture ⑩ was defined for stopping the motion of the mobile robot. Generally speaking, mobile robots used in complex auxiliary operation scenarios need to be equipped with mechanical arms or other actuators to complete the corresponding task operations. In order to realize the control priority switching between the mobile robot and its equipped actuator, gesture ⑨ was defined to achieve the switching of the control subject, thus completing the specified task smoothly.

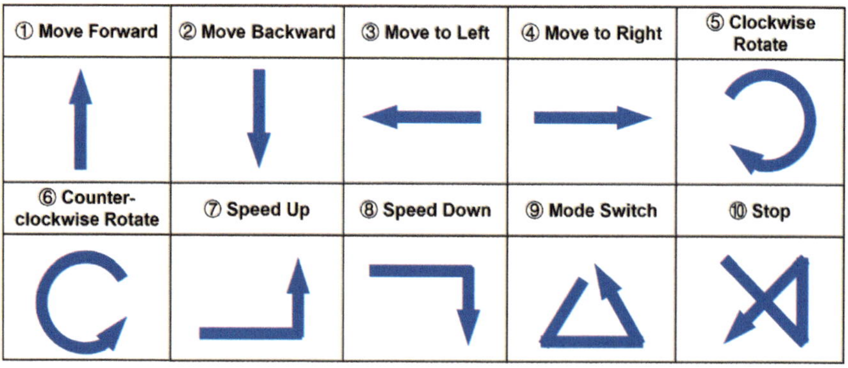

Fig. 4.7 Ten predefined gesture commands for mobile robot control, reprinted from Ref. [6], CC BY 4.0

4.2.2.1 Experimental Design and Performance Analysis of Dynamic Gesture Trajectory Recognition

In order to test the performance of the designed dynamic gesture trajectory recognition system, five subjects (including four males and one female in good health, aged between 20 and 30 years old) were selected as experimental subjects in this section. During the experiment, the five subjects completed gestures ①–⑩ as a set of gesture combinations according to the aforementioned predefined 10 gestures consecutively. Each subject repeated the above consecutive gesture combinations for 10 times in sitting and standing respectively. In this way, 200 gesture samples could be collected from each subject, and 500 gesture samples could be collected in sitting and standing respectively. Finally, a total of 1000 gesture samples were obtained. During the experiment, participants wore a wearable wrist camera on their right wrist and adjusted the position and Angle of the camera to ensure that the hand area occupied about 2/5 of the frame. After that, the operator made actions according to the 10 predefined gesture trajectory instructions to collect gesture trajectory data. Before the start of a single action, according to the above definition of gesture segmentation rules, the operator first made a wrist swinging action, and then made a single gesture trajectory. After completing a single gesture, the operator needed to maintain a static state for more than half a second. Under the guidance of the above rules, the subjects pre-collected a combination of gestures to ensure the validity of gesture data, which was not included in the collected 1000 experimental samples. After completing the acquisition and segmentation of all dynamic gesture trajectories and extracting the velocity data of valid individual gesture trajectories, the DTW algorithm was used to compute the variability of the gesture trajectories to classify the gestures and cross-validate the recognition accuracy.

(1) **Segmentation results of continuous gesture trajectory**:

According to the continuous gesture trajectory segmentation algorithm defined in the previous section of this chapter, all 100 sets of continuous gesture trajectories (10 sets of continuous gesture trajectories for each of the 5 subjects in sitting and standing positions) were segmented in order to extract valid individual gestures. The experimental results showed that 94 of these sets of continuous gestures were correctly segmented into 10 correct individual gesture intervals, and six sets of continuous gestures were not correctly segmented. Among the 6 wrongly segmented gestures, three of which failed due to incorrect identification of the start and end points because of the subjects' gesture irregularities (including wrist oscillation amplitude required before a single valid gesture being too small, and failing to maintain a stationary state after completing a single valid gesture). The remaining three sets of segmentation errors were caused by the operator's wrist swing during a single gesture, which make the hand region leave the frame, and resulted in a redundant number of segmentations. Finally, out of 1000 correct gestures, 992 gestures were completely segmented, with a segmentation accuracy of 99.20%. Individual gesture trajectories that were not correctly segmented were subsequently processed for manual segmentation by specifying the start and end points of the gestures.

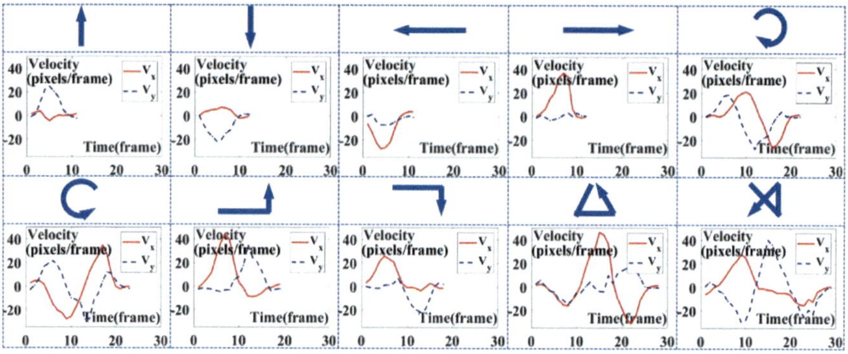

Fig. 4.8 Velocity profiles for ten predefined gesture commands, reprinted from Ref. [6], CC BY 4.0

(2) **Analysis of the results of background velocity calculations**:

According to the calculation results of gesture segmentation and background velocity based on SURF feature point matching, the velocity profile of a typical set of 10 predefined gestures in the experiment was selected and plotted, as shown in Fig. 4.8. From the figure, it can be seen that simple gesture trajectories such as gesture ①, ②, ③, and ④ have the shortest duration, with an effective gesture action duration of about 1 s, while complex trajectories such as gesture ⑤, ⑥, ⑦, ⑧, ⑨ and ⑩ have a duration of about 3 s. In order to reduce the influence of different gesture duration on classification, all gesture data were normalized using linear interpolation method.

(3) **Analysis of gesture instruction recognition results**:

After obtaining the valid velocity data of the corresponding gesture trajectories, the distance between the corresponding gesture velocity sequence data was calculated based on the DTW algorithm to classify the 1000 gesture samples. Three cross-validation methods were used as described below: (1) Leave-one-subject-out (LOSO), which selects one subject's sample data as the test set and other subjects' sample data as the training set; (2) Leave-other-subject-out (LPO), which selects one subject's sample data as the training set and selects other subjects' sample data as the test set; (3) Leave-one-group-out within one subject (LOOWS), which selects a group of samples from one subject as the test set and selects other sample groups of that subject as the training set. Based on the DTW algorithm, the type of gesture in the test set was determined as the recognition result based on the minimum DTW distance comparing the gesture trajectories of the training set.

The accuracy rates of gesture recognition using different cross-validation methods are listed in Table. 4.2. The average recognition accuracy under LOSO cross-validation achieved 97.34%, validating the performance of the system for unfamiliar operator gesture recognition. LPO cross-validation achieved an average accuracy of 96.55%, which is lower than that of LOSO, reflecting the differences between subjects and the diversity of the collected data samples. The LOOWS method achieved an average accuracy of 98%, which is higher than LOSO, indicating that operators can add their own characteristic gestures to the system and can recognize gestures effectively. As Table. 4.2 shows, the dynamic gesture recognition accuracy in the standing state is slightly higher than the gesture recognition accuracy in the operator's sitting state. In addition to the random influence from the external environment, the features of gestures in the standing state are more obvious than those in the sitting state, because the operator's movement space in the standing state is much larger than that in the sitting state.

In order to further validate the performance of the gesture recognition application for mobile robot interaction control, the gesture velocity data collected in the standing and sitting states under the LOSO cross-validation method were analyzed respectively. The confusion matrix of recognition results under the two gesture operation states was calculated, as shown in Fig. 4.9. Figure 4.9a on the left shows the confusion matrix of 10 gesture recognition results in the seated state. Figure 4.9b on the right shows the confusion matrix of 10 gesture recognition results in the standing state. In addition, based on the confusion matrix, the accuracy rate, recall rate, and F1 values of 10 gesture classifications were calculated to further verify the performance of gesture recognition, as shown in Fig. 4.10. Figure 4.10a represents the results in the sitting state. Figure 4.10b represents the results in the standing condition. The mean value of F1 value is as high as 0.984 in the sitting condition and 0.963 in the standing condition, indicating that its classification performance is better in the sitting condition than in the standing condition.

The confusion matrices in Fig. 4.9a and b respectively show the confusion results of 10 gestures under LOSO cross verification in sitting and standing states. Among the confused gestures, some kinds of gestures accounted for a slightly higher proportion of confusion. Now, this phenomenon is analyzed: gesture ⑦ is incorrectly identified as gesture ④, and the confusion ratio is 4.67% in the sitting state and 3.11% in the standing state, which is caused by the small longitudinal motion amplitude of some operators' gestures ⑦ along the y direction. For similar reasons, gesture ⑧ was mistakenly identified as gesture ④, with an error ratio of 4.67%. In addition, part of gesture ⑩ was incorrectly recognized as gesture ③ with an error ratio of 2.67%, which was caused by the fact that cross gesture ⑩ was recognized as a gesture segmentation by wrist swing in the front part of the gesture trajectory. In addition, there are also some gestures ⑧ that are incorrectly recognized as gesture ②, with a confusion ratio of 2.67%, which is caused by the small amplitude of the lateral movement along the x-direction of gesture ⑧.

4.2.2.2 Verification of Gesture Recognition Interaction Functions for Robot Mobility Control

After the validation experiments of the gesture trajectory recognition performance of the wearable wrist camera were completed, the practical application performance of this gesture track recognition scheme in the human–computer interaction system was further tested. According to the predefined 10 gesture trajectory commands, this

Table 4.2 Accuracy of gesture recognition under three different cross-validation methods

Operation status		Sitting state						Standing state						Mean
Subject number		1	2	3	4	5	Mean	1	2	3	4	5	Mean	
Recognize accuracy (%)	LOSO	97.33	96.78	93.68	97.89	95.67	96.27	99.67	98.17	96.44	99.33	98.48	98.42	97.34
	LPO	96.56	97.56	93.78	97.33	94.67	95.98	99.37	97.56	94.44	98.44	97.56	97.47	96.55
	LOOWS	97.67	90.00	100	100	100	97.53	98.33	100	94.00	100	100	98.47	98.00

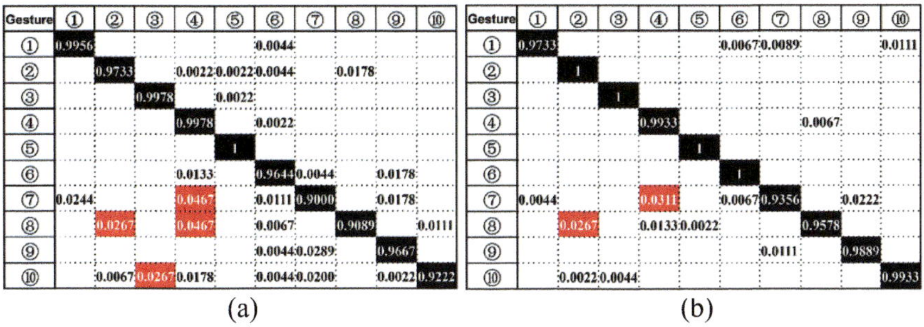

Fig. 4.9 Recognition confusion matrixes of the ten gestures, reprinted from Ref. [6], CC BY 4.0. **a** Confusion matrix in sitting state. **b** Confusion matrix in standing state

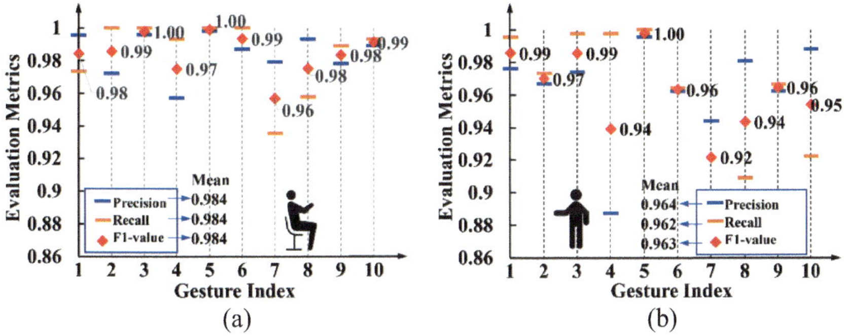

Fig. 4.10 Accuracy, recall and F1-score of gesture recognition, reprinted from Ref. [6], CC BY 4.0. **a** Recognition results in sitting state. **b** Recognition results in standing state

book applied the designed gesture recognition system to the control of the omnidirectional mobile chassis of a nursing robot for validation, and the selected nursing robot is a dual-arm robot platform built by autonomous integration for safe elderly assistance. During the functional verification experiments, the interaction performance of the nursing auxiliary robot was verified in two motion control modes: ride mode (operator sitting) and teleoperation mode (operator standing). The operator wears a wrist camera and makes a corresponding gesture for recognition. Unlike the experimental test and analysis part of the recognition performance in this chapter, the gesture recognition in this section of the functional verification process is performed online in real time, with classification based on a specified training set and real-time output of the gesture recognition results.

As shown in Fig. 4.11a, in ride-on mode, the user wore a wrist camera on his right hand and sat on the ride section in front of the mobile nursing robot. The operator gave control commands to operate the corresponding movements of the robot. In this mode, the user with lower limb disability can use the upper limb to make gesture commands to control the nursing auxiliary robot to make corresponding

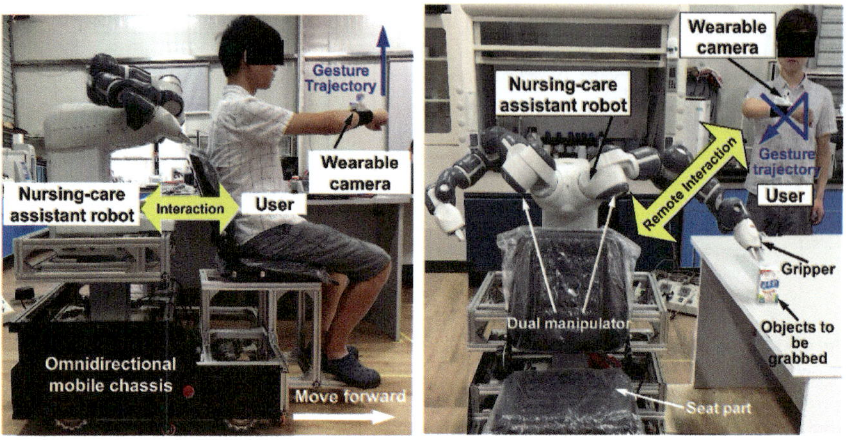

(a) (b)

Fig. 4.11 Dynamic gesture recognition for mobile care robot control, reprinted from Ref. [6], CC BY 4.0. **a** Validation of the riding operation mode. **b** Validation of teleoperation mode

auxiliary movements. Since the operator's own motion will be superimposed on the robot's motion when riding on the robot, the accuracy of gesture recognition is not significantly affected by the use of normalized velocity sequence features during the actual verification process, which demonstrates the robustness of the gesture trajectory recognition system and the corresponding algorithms proposed in this chapter. The gesture trajectory recognition in the operator's standing posture can be used in teleoperation scenarios of mobile care robots, as shown in Fig. 4.11b. The operator can make specified control commands based on predefined gestures and switch the control between the mobile robot and the upper robotic arm. During the application validation in this section, the operator's hand image data and the robot controller were connected via a Wi-Fi LAN, which was used to transmit the image data in time to the upper computer for recognition and processing, and sent it to the mobile robot for motion control.

4.3 Human-Motion Based Teleoperation of Robot Articulated Torso

4.3.1 A Dual-Arm Mobile Robot Prototype with Articulated Torso

In order to adapt to the flexible task requirements in nursing assistance scenarios, this book designed a humanoid two-armed nursing assistance robot with a 3-degree-of-freedom torso based on the Kinova Jaco2 collaborative robotic arm [13] and Bulldog wheeled mobile chassis, which can change the torso angle through torso motors in different task scenarios, and adjusted the workspace of the robot arm it supports to adapt to the unstructured task environment, as shown in Fig. 4.12.

There is a four-wheel differential moving chassis under the two-arm humanoid robot, which is mounted with a humanoid mechanical torso and Jaco2 twin mechanical arms. The humanoid torso consists of three servomotors, which can achieve torso movements in three directions: elevation, pitch, and rotation. During the tasks operation of the robot, the working space of the dual robotic arms can be adjusted by changing the lifting height and posture of the torso to ensure the flexibility of the end movement of the robotic arms. The rotating mechanism is driven by servo motor 1 to move the gear set, which can realize the symmetrical rotation in the horizontal direction within the range of 330°. The tilting mechanism is driven by servo motor 2 to drive the worm gear transmission to be responsible for the tilting action of the torso, which can achieve the motion range of 65° forward torso and 80° backward body. The lifting mechanism driven by the servo motor 3 can drive the ball screw to realize the rise and fall of the robotic arm within the range of 20 cm in the vertical direction.

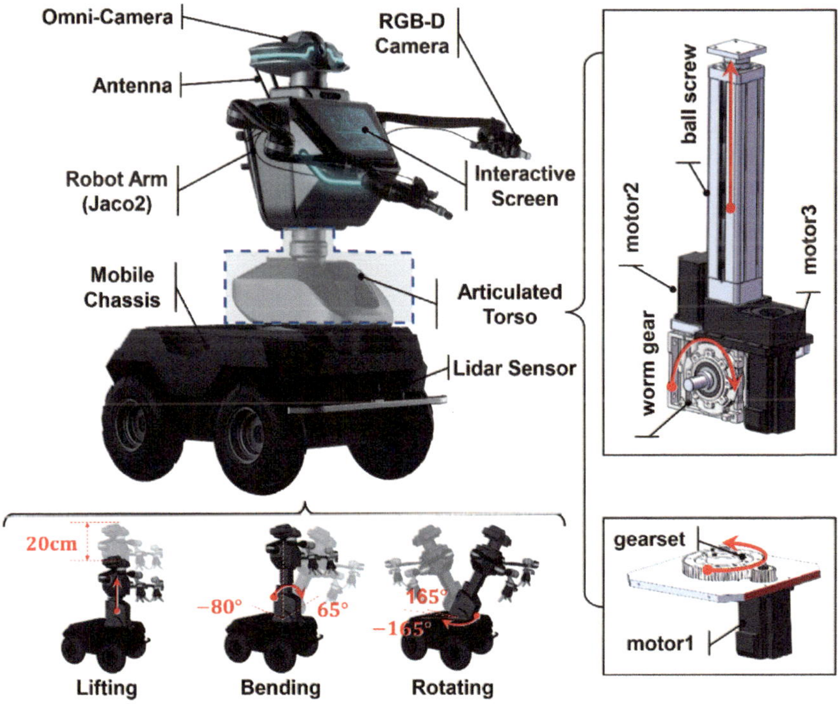

Fig. 4.12 The components overview of the anthropomorphic dual-arm robot

The mobile chassis of the robot is equipped with a LiDAR sensor Rapid A1 from Silan Technology, which is used for environmental perception and obstacle avoidance navigation during movement. The robot's upper arms are two Kinova Jaco2 collaborative arms with 6 degrees of freedom, a single arm extension of 98.4 cm, and a maximum load of 2.6 kg. The arms are equipped with an optional three-finger gripper to improve flexibility and efficiency in gripping the target object, with a maximum gripping force of 40 N and a maximum opening distance of 175 mm. The end of each Jaco 2 arm is equipped with an RGB-D camera RealSense [14] for visual feedback during the gripping operation. In addition, the humanoid robot is equipped with a pan-tilt camera on the top, which allows the main operator to observe the surrounding environment during teleoperation. To communicate with remote computer for debugging and remote operation, a high-power signal receiver is mounted on the back of the robot.

4.3.2 Motion Mapping Strategy for Robot Articulated Torso

As shown in Fig. 4.13, in this section, in order to realize the mapping between the operator's limb movements and the robot's torso motion pose, a wearable inertial motion capture device PN was adopted to capture the operator's limb movements, similar to the upper limb pose mapping teleoperation technology route. The operator's movement nodes are shown in Fig. 4.13a. To realize the mapping of the operating robot's rotation, pitching, and lifting actions through the operator's motion characteristics during teleoperation, the structural characteristics of the robot are compared with the human body's motion node coordinate system characteristics as shown in Fig. 4.13b.

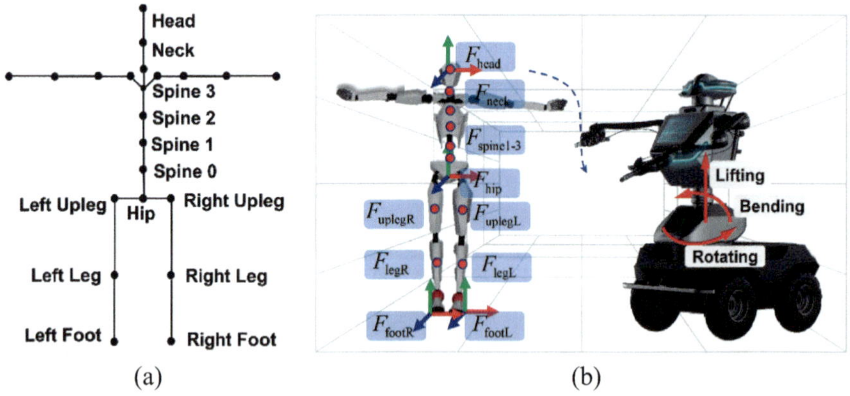

Fig. 4.13 Torso DoFs of the anthropomorphic robot compared to the operator. **a** Human skeletal nodes. **b** Human-robot torso motion mapping

After comparing the torso freedom features of the robot with the movement node features of the operator's body, the pose mapping rules for controlling the three torso movements of the robot were designed, as shown in Fig. 4.14. The design of the lifting and lowering bit-posture mapping rules for the robot torso is shown in Fig. 4.14a. In the human body movement behavior, the integrated selection of the right foot and the left foot node relative to the hip distance δz_{leg} is the mapping input of the lifting action. During the control of the lifting action of the robot, the operator can make a standing up action or squatting action to achieve the change of the distance Z, mapping the position control of the servo motor 3 input to the torso lifting mechanism, and realize the position and pose mapping of the lifting action. The design of the pitch pose mapping rule for the robot torso is shown in Fig. 4.14b. Since the spine 3 nodes of the operator will rotate along the Y-axis relative to the hip node during the pitching process, the position control of the servomotor 2 of the pitching mechanism is realized by judging the Euler angle transformation of the spine 3 nodes of the operator relative to the hip node along the Y-axis Euy_{spine}, realizing the bit-pose mapping of the pitching action. Finally, as shown in Fig. 4.14c, the most obvious change characteristic of the operator during the turning process is the angular change of the spine 3 node relative to the hip node along the Z-axis. Therefore, the rotational action mapping of the robot torso is controlled by the positional control of the servo motor 1 in the rotational mechanism, which is realized by judging the rotational change of the spine 3 node relative to the hip node along the Z-axis Euz_{spine},.

4.3.3 Validation of Human-Motion Based Robot Torso Control

Based on the above-mentioned torso motion pose mapping strategy, the designed redundant degree-of-freedom humanoid robot was verified in the laboratory for the torso pose mapping teleoperation function. The operator wore a wearable inertial motion capture device PN to make the corresponding movements, as shown in Fig. 4.15. With the ROS control framework built above, the operator's whole-body node coordinate data was mapped to the robot controller. After receiving the topic, the industrial computer on the robot mapped the output control signals of the motor drivers of the three servo motors in the torso to make corresponding position control for each motor. As shown in Fig. 4.15a, when the operator made a squatting action, the distance from the left and right feet to the hip nodes changed, and the robotic arm lifting mechanism descended. As shown in Fig. 4.15b, when the operator made a prone action, the operator's spine 3 node relative to the hip node along the Y-axis changed in angle, with the mapping control torso tilt mechanism tilting forward, and the robot made a prone action. As shown in Fig. 4.15c, when the operator made a turning action, the operator's spine 3 node relative to the hip node along the Z-axis changed the angle, and the mapping controlled torso rotation mechanism along the horizontal plane to rotate toward left or right, with the robot making a rotation

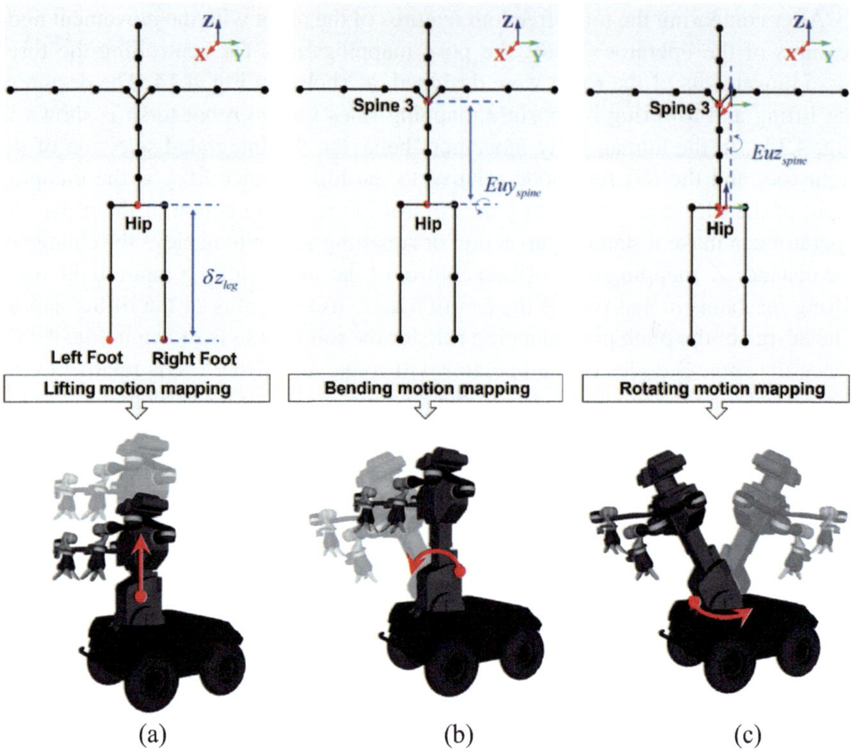

Fig. 4.14 Motion mapping strategy for 3-DoF anthropomorphic robot torso. **a** Lifting motion mapping. **b** Bending motion mapping. **c** Rotating motion mapping

action. So far, the remote operation application verification of pose mapping for the multi-degree-of-freedom torso of the humanoid robot has been completed, and the feasibility and practicability of the pose mapping strategy for the torso proposed in this book has been verified in the process.

4.4 Summary

Robots oriented to unstructured application scenarios such as homecare or medical assistance are generally responsible for controlling the robot's movement and robotic arm workspace adjustment by moving the chassis and other redundant degree-of-freedom proprioceptive motion mechanisms. After realizing single-arm and dual-arm pose mapping teleoperation based on upper limb movements, in order to realize teleoperation control of nursing assistance robots, this chapter carried out an intuitive teleoperation control study based on operator limb motion capture for the chassis

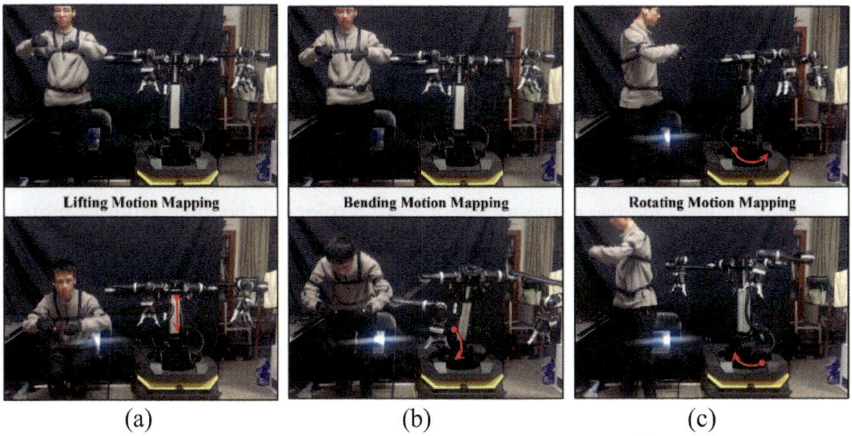

Fig. 4.15 Practical teleoperation of a 3-DoF anthropomorphic robot torso. **a** Lifting motion mapping. **b** Bending motion mapping. **c** Rotating motion mapping

movement control and the pose mapping control of the humanoid torso, respectively. A dynamic trajectory recognition method of operator end gesture based on background velocity feature matching was designed for the chassis movement teleoperation control of the robot, and the chassis movement control of the nursing assistive robot was realized by a wearable wrist camera based on the recognition results of multiple gesture commands. For the teleoperation control of the redundant-degree-of-freedom humanoid torso of the nursing care assistive robot, this chapter designed a torso pose mapping teleoperation control method for the slave robot based on the torso movement and lower limb movement data of the operator at the master end. Further application studies of the proposed bit-posture mapping teleoperation technique were carried out for two types of application scenarios, namely remote homecare and telemedical assistance. The main conclusions of this chapter are as follows:

(1) A method for identifying dynamic gesture trajectories at the operator's end was proposed. By extracting feature points in the background of the wrist camera worn by the operator to analyze the operator's movements, it solved the problem of the current common visual recognition solution that the operator is restricted by the layout of visual equipment and fixed operating space, and improved accuracy and flexibility of dynamic gesture trajectory recognition. 1000 gesture samples were collected and analyzed for 10 types of gesture trajectories. The average recognition accuracy under LOSO cross-validation reached 97.34%. Furthermore, mobile control gesture instructions were designed based on the implemented gesture trajectory recognition system, and the application verification of mobile control of nursing assistance robots was realized.

(2) Facing the demand for torso motion pose mapping teleoperation control of multi-degree-of-freedom humanoid robots, this chapter proposed a multi-degree-of-freedom torso pose mapping teleoperation method based on the characteristics of the operator's motion at the main end. Based on the relative coordinate transformation between the coordinates of the hip joint nodes and the foot nodes, and the transformation between the coordinates of the hip joint nodes and the spine nodes, the pose mapping teleoperation rules for the three kinds of motion control of rotation, pitch and lift of the multi-degree-of-freedom humanoid robot were designed. Moreover, the functional verification was implemented on an integrally constructed solid robot, which verified the practicability of the torso motion pose mapping teleoperation control method proposed in this chapter.

(3) Based on the research on pose mapping teleoperation technology, this chapter integrated the design of an auxiliary robot for the elderly in the home environment based on the needs of remote homecare scenarios. Moreover, the application verification of remote teleoperation to assist in grabbing and delivering items was carried out to meet the daily needs of the elderly. In response to the demand for remote medical assistance, a teleoperated medical assistance robot system for the COVID-19 isolation ward was integrated and built. This study received clinical ethics review approval and carried out clinical functional verification in the isolation ward of the First Affiliated Hospital of Zhejiang University School of Medicine during the epidemic, validating the practical performance of the posture mapping intuitive teleoperation technology studied in this article.

References

1. Sereinig M, Werth W, Faller L-M (2020) A review of the challenges in mobile manipulation: systems design and RoboCup challenges. Elektrotechnik und Informationstechnik 137(6):297–308
2. Štibinger P, Broughton G, Majer F et al (2021) Mobile manipulator for autonomous localization, grasping and precise placement of construction material in a semi-structured environment. IEEE Robot Autom Lett 6(2):2595–2602
3. Chen F, Lv H, Pang Z et al (2019) WristCam: a wearable sensor for hand trajectory gesture recognition and intelligent human-robot interaction. IEEE Sens J 19(19):8441–8451
4. Avalos J, Cortez S, Vasquez K et al (2016) Telepresence using the Kinect sensor and the NAO robot. In: IEEE Latin American symposium on circuits and systems, pp 303–306
5. Siaw T, Han Y, Wong K (2023) A low-cost marker-based optical motion capture system to validate inertial measurement units. IEEE Sens Lett 7(2):1–4
6. Yang G, Lv H, Chen F et al (2018) A novel gesture recognition system for intelligent interaction with a nursing-care assistant robot. Appl Sci-Basel 8(12):2349
7. Ai SJ, Wang X, Ma MC et al (2013) A method for correcting non-linear geometric distortion in ultra-wide-angle imaging system. Optik 124(24):7014–7021
8. Achanta R, Shaji A, Smith K et al (2012) SLIC superpixels compared to state-of-the-art superpixel methods. IEEE Trans Pattern Anal Mach Intell 34(11):2274–2281
9. Li Y, Sun J, Tang C et al (2004) Lazy snapping. ACM Trans Graph 23(3):303–308
10. Li J, Wang Y, Wang Y (2012) Visual tracking and learning using speeded up robust features. Pattern Recogn Lett 33(16):2094–2101

11. Lowe DG (2004) Distinctive image features from scale-invariant keypoints. Int J Comput Vis 60(2):91–110
12. Paliwal KK, Agarwal A, Sinha SS (1982) A modification over Sakoe and Chiba's dynamic time warping algorithm for isolated word recognition. Signal Process 4(4):329–333
13. Golluccio G, Gillini G, Marino A et al (2020) Robot dynamics identification: a reproducible comparison with experiments on the Kinova Jaco. IEEE Robot Autom Mag 28(3):128–140
14. Tadic V, Odry A, Kecskes I et al (2019) Application of intel realsense cameras for depth image generation in robotics. WSEAS Trans Comput 18:2224–2872

Chapter 5
Communication in Human-Motion Based Robot Teleoperation

Abstract This chapter explores the impact of wireless network performance on the control of robotic teleoperation systems, particularly in terms of pose-mapping teleoperation. With the advancement of information and network technology, wireless networks, including private 5G and Wi-Fi 6, have increased the flexibility and cost-efficiency of robotic systems by simplifying design, installation, and maintenance. However, these networks also introduce transmission delays and communication instability, which can significantly affect the time-sensitive nature of robotic teleoperation. The chapter systematically evaluates the performance of various local wireless networks and examines how these delays influence control performance, discussing network co-design strategies to optimize performance for different requirements. Furthermore, it proposes a cloud server-based virtual network framework for ultra-long-distance, intercontinental teleoperation, enabling human-robot pose mapping over 7800 km from Sweden to China. To mitigate the challenges of motion errors and delays caused by wide-area networks, a feedforward controller is introduced to improve teleoperation control. The proposed framework and controller are verified using a Kinova dual-arm robot platform, demonstrating effective teleoperation over vast distances and improving control stability despite long-distance connectivity issues.

Keywords Wireless networks · 5G · Wi-Fi 6 · Intercontinental teleoperation · Feedforward controller

5.1 Introduction

With the development of information technology and network technology, wireless networks have made robotic systems more flexible, simplified the design and installation process, and reduced the maintenance costs of wired network deployments. Meanwhile, wide-area network connectivity has made it possible for operators to remotely teleoperate robots over different distances. Local wireless networks

H. Lyu et al., *Human Motion Awareness and Robot Teleoperation*, Advanced Topics in Science and Technology in China 4, https://doi.org/10.1007/978-981-96-6545-7_5

and wide-area wired networks have made robotic teleoperation systems more flexible and extended the implementation distance of teleoperation systems, making them an indispensable part of teleoperation systems. At the same time, wireless networks introduce additional transmission delays and unreliable communications, which has a significant impact on time-sensitive robotic teleoperation control, and long-distance wide-area networks also pose challenges to teleoperation systems in terms of extremely large delays and control stability [1]. This chapter systematically investigates the impact of wireless local area network performance on the control performance of robotic teleoperation. Long-term performance testing and evaluation were conducted on various types of local wireless networks, including private 5G and Wi-Fi 6, with respect to the control performance of the aforementioned pose-mapping teleoperation system. Different network co-design methods for various control performance requirements were discussed. Furthermore, a cloud server-based virtual network intercontinental ultra-long-distance human–robot pose-mapping teleoperation framework was proposed. The study examined the impact of intercontinental ultra-long-distance wide-area network connectivity performance on teleoperation control systems, and achieved teleoperation over 7800 km from Sweden to China across the Eurasian continent. To address significant motion errors and delays between the master and slave ends, a feedforward controller was designed to enhance teleoperation control performance under intercontinental ultra-long-distance wide-area network connectivity, and intercontinental ultra-long-distance pose-mapping teleoperation functionality verification was implemented on the Kinova dual-arm robot platform.

5.2 Robot Teleoperation via Local Wireless Networks

5.2.1 Performance Evaluation Under Local Area Network Conditions

Traditional methods typically rely on empirical model-based simplifications of network channel modeling to assess the impact of wireless networks on control systems, which often fail to replicate the uncertain events encountered in real wireless network connections and cannot accurately evaluate network reliability. In order to systematically quantify and evaluate the network performance and control performance of teleoperation control systems under local area network connection conditions, this book designs a semi-physical simulation test system based on network hardware-in-the-loop. The architecture of this system is illustrated in Fig. 5.1, consists of four components: a robot controller simulation platform, a robot actuator simulation platform, network monitoring devices, and deployment in a real network environment.

Among which, network monitoring devices employ the Bifrost ET2000 Ethernet multi-channel probe [3], which can listen to inter-device communication in the

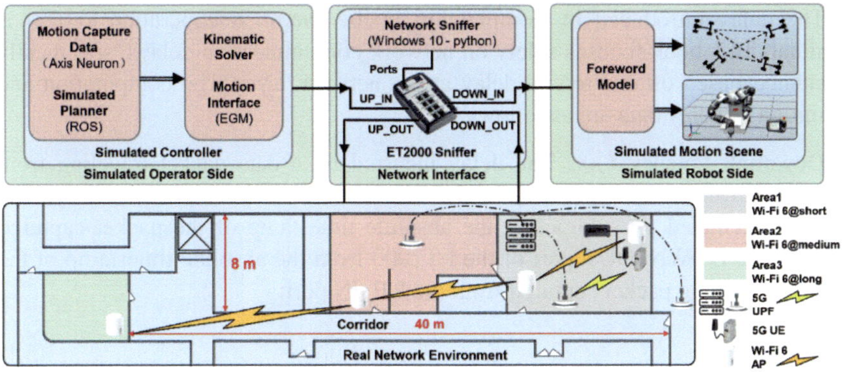

Fig. 5.1 Diagram of the network hardware-in-the-loop framework, reprinted from Ref. [2], CC BY 4.0

network without introducing additional latency. The simulation controller and actuator of the robot are connected through a real network test environment and accessed through the network listening interface. The controller is responsible for preprocessing pose data for remote operation and kinematic calculations of the robot, it also converts and outputs joint space data for the EGM interface in real-time and at a high frequency. The written network listening program continuously records real-time raw data packets from the controller port and actuator port, and calculates the packet latency, as shown in Fig. 5.2

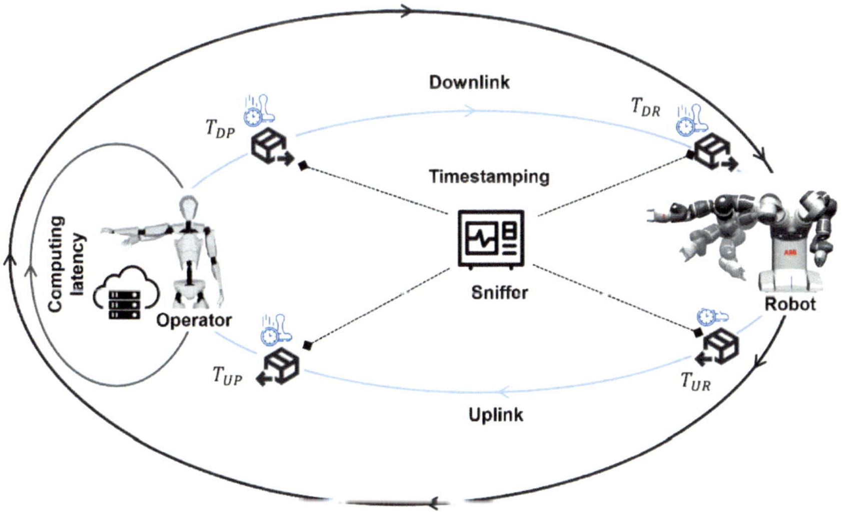

Fig. 5.2 Network latency measurement using the ULT timestamp device

The delay introduced by computation in the overall teleoperation system is minimal. This book focuses solely on network communication delays, specifically calculated based on end-to-end delay at the network layer. The computation and definition of delay data are as follows:

- Downlink Latency L_D : The delay incurred by a data packet traveling from the controller end (operator) to the actuator end (robot) within the system. It is calculated by subtracting the absolute timestamp of the packet captured from the DOWN_OUT port of the ET2000 from the absolute timestamp of the corresponding packet captured from the UP_IN port:

$$L_D = T_{DR} - T_{DP} \tag{5.1}$$

- Uplink Latency L_U : The delay generated within the system by a data packet traveling from the actuator end (robot) to the controller end (operator). It is calculated by subtracting the absolute timestamp of the packet captured from the DOWN_IN port of the ET2000 from the absolute timestamp of the corresponding packet captured from the UP_OUT port.

$$L_U = T_{UP} - T_{UR} \tag{5.2}$$

During practical testing, the Wireless Local Area Network (Wi-Fi 6) access points (APs), private 5G base stations (UPF, User Plane Function), and user equipment (UE) were strategically deployed as depicted in the floor plan diagram shown in Fig. 5.1. For Wi-Fi 6 network evaluations, the distances between two AP nodes were configured across three distinct environments: close-range, mid-range, and long-range (referred to as Close-range Wi-Fi 6, Mid-range Wi-Fi 6, and Far-range Wi-Fi 6, respectively). The private 5G network, sourced from Ericsson's private industrial wireless solution, facilitated connectivity between the 5G terminal access device (UE) and the 5G base station gateway (UPF). Utilizing Phoenix Contact's industrial-grade cellular router node, the 5G terminal access device was selected for this study. Ethernet connectivity was established through cable connections between the DOWN_OUT and UP_OUT ports of the ET2000. To ensure uninterrupted monitoring across various wireless local area networks, each Wi-Fi 6 AP node was linked to the DOWN_OUT and UP_OUT ports, while the base station gateway (UPF) of the 5G network was connected to the DOWN_OUT port, and the UE was connected to the UP_OUT port.

5.2.1.1 Network Configuration and Experimental Setup

As shown in Fig. 5.3, the operator utilizes motion capture devices to capture human body motion data, which undergoes transformation and transmission to the controller at the robot's end using the aforementioned pose mapping strategy to control the YuMi robot. This system is based on ABB's EGM Position Guidance option interface. EGM

Position Guidance is specifically designed for advanced robot users, offering a low-level position guidance control interface that bypasses relevant path planning. Users can swiftly read positions of the relevant motion system and write positions to the system at high speeds, with a maximum frequency of 250 Hz (up to every 4 ms). Compared to other external motion control methods, EGM Position Guidance boasts advantages such as higher speed, lower loop delay, and reduced wait time. While EGM typically accepts input data in the form of joint values or end-effector pose values for motion tracking and guidance, in the case of 7-axis robots like the YuMi dual-arm robot utilized in this system, only joint values can be utilized as guidance data input. This limitation stems from the inability of redundant freedom robotic arms to perform high-frequency inverse kinematics calculations within the controller and output them in real-time. Therefore, as illustrated in Fig. 5.3, the processing of pose data at the operator end is achieved through pre-computed inverse kinematics within the master controller. The focus of this chapter's work primarily revolves around the impact of wireless communication links on the transmission of joint space data calculated at the operator end and sent for control of the slave-side robotic arm.

Firstly, real-time motion data of the operator is captured using Axis Neuron software running on a Windows system. A Ros_Serial communication interface is developed based on the NeuronDataReader SDK to transmit the human motion data from the Windows end to the Ubuntu system. At the Ubuntu end, the motion control program retrieves the joint coordinate data of the human body and processes it using the pose mapping strategy designed in this book to convert it into the end-tool center point data of the robot. Subsequently, real-time calculation of the robot's

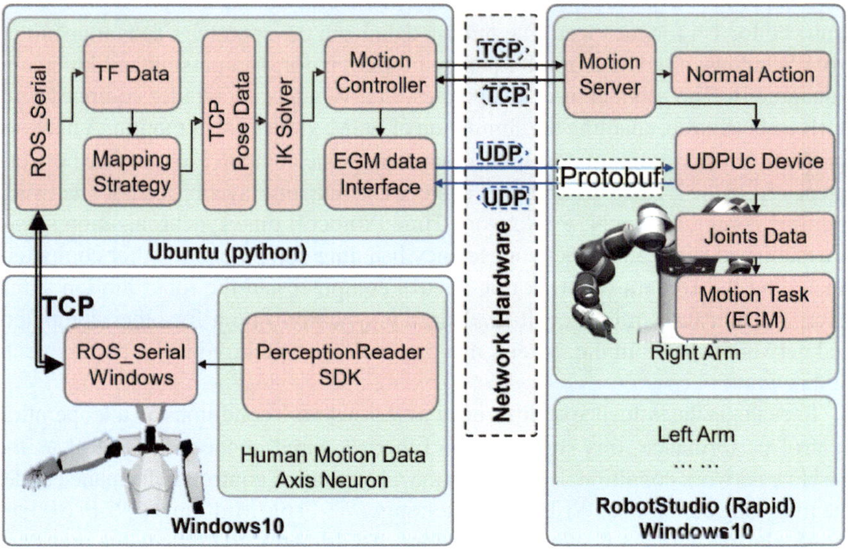

Fig. 5.3 Network configuration and data flow of the teleoperation system, reprinted from Ref. [2], CC BY 4.0

desired joint space position data is performed using an inverse kinematics solver. The motion control program establishes a connection with the robot-side motion control processing program via the TCP protocol. The motion control processing program on the robot side controls the general motion of the robot, such as linear and joint motion, as well as other advanced functionality interfaces and external I/Os, based on the given robot target positions. EGM control is enabled through pre-established TCP communication (based on ABB's PC Interface 616-1), establishing a data connection between the master and slave EGM interfaces. The EGM protocol utilizes Google Protocol Buffers for encoding, offering advantages in terms of transmission speed and language neutrality. Given that the transmitted data consists of high-frequency real-time data, a certain packet loss rate is tolerable. According to the Google Protocol Buffers protocol, the robot side sends robot status data to the master side (operator side) or reads command data from the master side every 4 ms, regardless of when the sensor sends data.

This section's comparative experimental design aims to evaluate how the introduction of wireless networks affects robot control performance, while also examining the correlation between control performance parameters and communication performance parameters from a control perspective. By the previously outlined experimental system configuration, a computer running the motion capture software Neuron Axis is utilized to capture and parse human motion data. The operator wears a wearable inertial motion capture device and performs repetitive swinging motions with their right arm along the Y-axis direction of the global coordinate system, repeated eight times. These motions are then translated into control signals for the robot on the slave side. To ensure consistency in the motion data from the master side during the comparative experiment, the operator's human motion data is pre-recorded and replayed for 1 h for each test. The motion data from the operator side is transmitted from Windows to Ubuntu at a frequency of 125 Hz for data parsing and kinematic calculations. The robot controller operates on a Windows computer equipped with ABB RobotStudio, enabling the simulation of EGM-guided robot motion. A network loopback latency program runs on another standalone computer to measure network latency from the ET2000 detector. To achieve absolute time synchronization between multiple computers, an NTP (Network Time Protocol) time synchronization server is established between the network latency listening computer and other computers (including the Ubuntu robot motion control computer and the robot motion simulator). This ensures millisecond-level timing synchronization for other computers and network devices in the system, meeting the time synchronization requirements of this work.

To evaluate the influence of different wireless network conditions on teleoperation control performance, this study selects Ethernet wired connection control as the baseline network condition. It compares the teleoperation control performance under the following conditions: Wi-Fi 6@Close-range, Wi-Fi 6@Mid-range, Wi-Fi 6@Far-range, and 5G wireless network connections. Additionally, to mitigate the impact of wireless local area networks on the control performance of the slave side, this section proposes a method for filtering EGM-received data on the robot's slave side to address oscillations in joint control caused by network condition fluctuations. A comparison

Table 5.1 Experiment setup under different network conditions

Network conditions	Wired connection Ethernet	Wireless connection Wi-Fi 6	Wireless connection Wi-Fi 6	Wireless connection Wi-Fi 6	Wireless connection 5G
		Short-range	Mid-range	Long-range	
No filtering	Baseline comparison	Performance comparison	Performance comparison	Performance comparison	Performance comparison
Filtering	Filtering comparison	Performance comparison	Performance comparison	Performance comparison	Performance comparison

is made between controlling the robot using raw joint control data from the master side and using filtered joint data. In summary, the comparative experimental design is outlined in Table 5.1.

5.2.1.2 Key Metrics Definition for Control and Communication Evaluation

To delve deeper into the relationship between robot teleoperation control performance and network communication, this section employs key error metrics for characterizing control system performance and network performance indicators to assess teleoperation performance under varying network conditions. Here, the desired joint position values during teleoperation (joint position data sent from the master side) are denoted as J_d, while the measured actual joint position values of the robot (real-time joint position feedback from the EGM interface) are represented by J_r. J_d and J_r signify time series signals recorded based on system timestamps at a sampling rate of 500 Hz.

During the back-and-forth motion of the operator, the position variation of joint 1 of the robotic arm resembles a sinusoidal-like signal. By subtracting the timestamps corresponding to the peaks of the J_d and J_r waveforms, we can determine the observed motion latency $\hat{\tau}$ between the master and slave ends, as expressed in Eq. (5.3). The motion latency $\hat{T}$ for the reciprocating motion cycle in a single test is determined by Eq. (5.4).

$$\hat{\tau}_i = \arg\max_{t \in N_i} J_r(t) - \arg\max_{t \in N_i} J_d(t) \tag{5.3}$$

$$\hat{T} = [\, \hat{\tau}_1 \; \hat{\tau}_2 \; \dots \; \hat{\tau}_P \,] \tag{5.4}$$

where $i = [1, 2, 3, \dots, P]$, N_i represents the ith back-and-forth motion cycle.

The cross-correlation between J_d and J_r, denoted by the cross-correlation coefficient $\hat{R}_{J_d J_r}$, is computed as per Eq. (5.5). The average motion latency T between the two position data sequences is calculated according to Eq. (5.6).

$$\hat{R}_{J_dJ_r}(m) = \begin{cases} \sum_{n=0}^{N-m-1} J_{dn}^* + mJ_{rn}^*, m \geq 0, \\ \hat{R}_{J_rJ_d}^*(-m), m < 0. \end{cases} \tag{5.5}$$

$$\overline{T} = \underset{m \in [-\frac{T}{2}, \frac{T}{2}]}{\arg\max} \hat{R}_{J_dJ_r}(m) \tag{5.6}$$

To further evaluate the performance of the remote teleoperation control system, the difference between the expected joint position J_d and the measured actual joint position J_r at the peaks of their respective waveforms is computed as the motion peak error ε_{peak}, as shown in Eq. (5.7). The sequence of peak errors E_{peak} during a single test process is represented by Eq. (5.8).

$$\hat{\varepsilon}_{peak_i} = \max_{t \in N_i} J_r(t) - \max_{t \in N_i} J_d(t) \tag{5.7}$$

$$\hat{\mathrm{E}}_{peak} = \left[\hat{\varepsilon}_{peak_1} \ \hat{\varepsilon}_{peak_2} \ \ldots \ \hat{\varepsilon}_{peak_3} \right] \tag{5.8}$$

Furthermore, the calculation of the peak average error $\overline{\mathrm{E}_{peak}}$ is given by Eq. (5.9):

$$\overline{\mathrm{E}_{peak}} = mean(\hat{\mathrm{E}}_{peak}) = \frac{\sum_1^P \hat{\varepsilon}_{peak_p}}{P} \tag{5.9}$$

To assess network performance, the latency L is recorded during each experiment, covering both uplink latency L_U and downlink latency L_D. Further analysis involves computing and examining the normalized probability density function (PDF) and complementary cumulative distribution function (CCDF) of latency L. This statistical approach provides insights into the overall performance of network conditions.

5.2.1.3 Intuitive Comparison of Control Performance Under 5G and Wi-Fi 6

According to the set network testing conditions described above, this section first conducts a comparative analysis of the performance of four wireless network conditions involved in this chapter: including three different distances of Wi-Fi 6 connections and 5G network connections. Using the previously mentioned network sniffing devices, it tests and records the uplink latency L_U and downlink latency L_D during teleoperation under these four local wireless network connection conditions in real-time. Subsequently, statistical analysis is performed to calculate and analyze the PDF and CCDF curves under each network condition from a statistical perspective. As illustrated in Fig. 5.4, the PDF and CCDF curves depict the performance of four local wireless network conditions: close-range Wi-Fi 6, mid-range Wi-Fi 6, far-range Wi-Fi 6, and 5G, from left to right. These curves serve to characterize network performance. Upon comparing the performance curves of Wi-Fi 6 and 5G conditions,

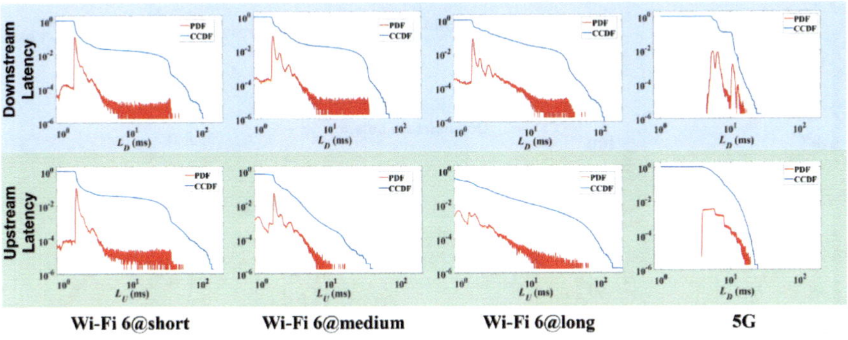

Fig. 5.4 The network profile analysis of Wi-Fi 6 and 5G network conditions, reprinted from Ref. [2], CC BY 4.0

distinct differences emerge: the CCDF curve of Wi-Fi 6 networks demonstrates a "large head, short tail" (LHST) characteristic, indicating a higher probability distribution of latencies above 10 ms. In contrast, the CCDF curve of 5G networks displays a "short head, large tail" (SHLT) feature, suggesting a lower probability distribution of latencies above 10 ms compared to Wi-Fi 6. These observations align with the inherent performance characteristics of the two wireless networks. The differences stem from the operational mechanisms of each network type. The flexibility of time-domain scheduling in 5G (NR, New Radio) systems results in more bounded network latencies. Conversely, Wi-Fi 6 network transmission, governed by the Carrier Sense Multiple Access with Collision Avoidance (CSMA/CA) algorithm, entails a waiting period for devices before packet transmission. This leads to a CCDF latency distribution curve of Wi-Fi 6 networks with a smaller head and a larger tail compared to 5G networks, indicating a higher occurrence of large latencies.

5.2.2 *Impacts of Local Wireless Networks to Teleoperation*

5.2.2.1 Impacts of Wireless Network Fluctuation on Teleoperation

To visually evaluate how changes in network connectivity environment affect the real-time pose mapping teleoperation system's control performance, this chapter introduces multiple AP connections within the same frequency band in the Wi-Fi 6@short-range network conditions scenario, as illustrated in Fig. 5.5. This subsection of testing is divided into two stages: Stage 1 and Stage 2. As depicted in Fig. 5.5b, the introduction of additional AP connections leads to a noticeable increase in both uplink latency L_U and downlink latency L_D during Stage 2, attributed to frequency interference. This is evident from the probability statistical curves of network latency, as illustrated in the PDF and CCDF curves in Fig. 5.5a and c, respectively. In Stage 2, both uplink latency L_U and downlink latency L_D exhibit a higher occurrence of

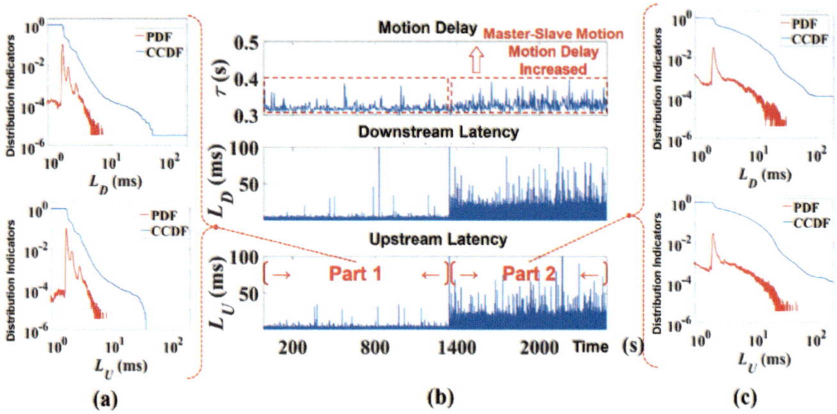

Fig. 5.5 The robot motion performance is affected by sudden changes in network conditions, reprinted from Ref. [2], CC BY 4.0

high-latency distribution, with longer "tails" in the CCDF curves, indicating higher latencies throughout the entire statistical period of Stage 2. Consequently, the motion latency τ at the master and slave ends during Stage 2 is notably higher compared to Stage 1. This underscores the significant impact of network performance fluctuations (resulting in increased network latency) on the teleoperation control performance of this system.

5.2.2.2 Statistical Analysis for the Impact of Wireless on Teleoperation

To comprehensively assess the influence of overall network performance on control system efficiency, this section initially employs wired Ethernet connection as the reference for teleoperation performance comparison experiments. It contrasts control performance parameters across four distinct wireless network scenarios: short-range Wi-Fi 6, medium-range Wi-Fi 6, long-range Wi-Fi 6, and 5G network environments, focusing on motion latency and peak error. Figure 5.6a delineates motion latency data curves for both wired and wireless network connections over a testing interval. Under the Ethernet wired connection, the master–slave motion latency stands at 0.3138 s. Despite wireless network conditions exhibiting only slightly higher overall motion latency with average latencies of 0.3229 s, 0.3140 s, 0.3213 s, and 0.3139 s for short, medium, and long-range Wi-Fi, and 5G networks respectively, the fluctuations in master–slave motion latency are conspicuous. This suggests the unstable nature of network loop connections under wireless conditions. Regarding joint motion peak error, depicted in Fig. 5.6b, the average peak error under wired connection conditions is −2.232°. It is evident that compared to wired connections, errors are more pronounced in the four wireless network environments. Moreover, peak motion error under the three Wi-Fi 6 wireless network conditions is slightly higher than that under 5G, with average error values of −2.841°, −2.851°, and −2.239° respectively for short, medium, and long-range Wi-Fi 6 conditions, while the average error value under the 5G network condition is −2.724°. Through the aforementioned data comparison, it is apparent that alterations in local network connection environments, particularly wireless setups versus wired ones, exert a notable impact on the overall control performance of teleoperation systems, particularly concerning peak joint motion error at the master–slave ends. Furthermore, the motion loop latency at the master–slave ends, which is approximately 300 ms, primarily pertains to control loop latency. Notably, the latency introduced by wireless networks, as opposed to wired networks, does not significantly affect control loop latency.

To further compare how overall network performance affects control efficiency, we conducted continuous motion testing under each network condition using the setup described earlier. Each network condition underwent a continuous, uninterrupted 4-h test to evaluate the master–slave motion tracking performance. Additionally, we calculated peak errors for both master–slave and joint motions. Each cycle of reciprocal motion at the operator end takes approximately 4.033 s to record. Thus, each test under each network condition yielded no fewer than 3600 performance parameter samples. Furthermore, using the aforementioned ET2000 network sniffer, we obtained no fewer than 2,000,000 network latency data samples continuously over a 4-h period.

The aforementioned sample data were collected in real-time and correspond to the system's absolute timestamps. Statistical analysis was performed on these control performance and network performance sample parameters. Similar to the analysis of network latency PDF and CCDF curves described earlier, probability distribution curves were plotted for control performance parameters under each test condition. As shown in Fig. 5.7, the PDF distributions of control performance parameters under four different wireless network conditions were compared with Ethernet wired connection. Figure 5.7a–d respectively illustrate the PDF curves of master–slave motion delay and joint motion peak error under Wi-Fi 6 networks at near, medium, and

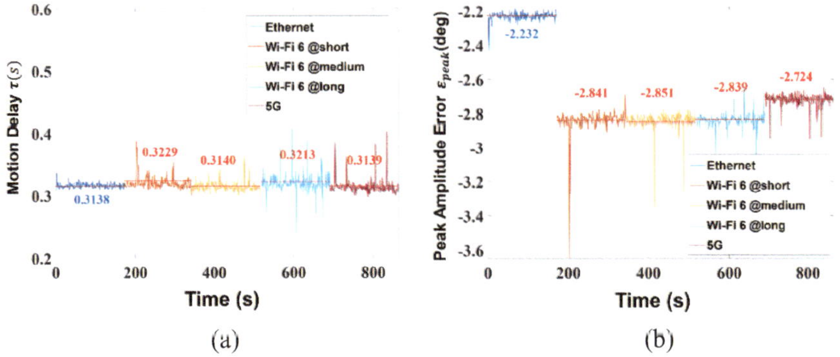

Fig. 5.6 Overview comparison of control performance using the raw data, reprinted from Ref. [2], CC BY 4.0. **a** Motion delay. **b** Joint position peak amplitude error

far distances, as well as 5G network, compared to wired Ethernet connection. It is evident that, compared to wired connection, the distribution of master–slave motion delay under four wireless network conditions is similarly concentrated, but wireless connections have wider distributions along the x-axis, with significant probability distribution between 0.35 s and 0.5 s. However, the distribution of joint motion error under wireless network conditions is notably different from wired connection. It is apparent that the distribution of joint motion peak error under four wireless network conditions is significantly larger than that under wired connection, with the peak error in Wi-Fi 6 networks at three distances being larger than that under 5G network conditions.

To mitigate the influence of wireless network fluctuations and communication uncertainties on teleoperation control performance, a low-pass filter was implemented at the receiving end of the robot's data on the slave side. This filter aimed to smooth out the received joint motion data, employing a 4th-order Butterworth design with a cutoff frequency of 250 Hz to ensure a balanced frequency response. After integrating the low-pass filter into the robot's system, comparative experiments were conducted under the same conditions as the previous wired Ethernet baseline, encompassing four distinct wireless network connection scenarios. The intuitive visual contrast of the experiment's outcomes is depicted in Fig. 5.8. When comparing the master–slave motion delay and joint peak error with and without the filter on the slave side, as illustrated in Fig. 5.6, a notable observation emerges. The introduction of the low-pass filter markedly reduces joint peak errors under wireless local area network conditions, effectively aligning them with the performance observed under wired connections. However, this improvement comes at a slight cost: the introduction of the filter marginally increases the motion delay of the master–slave system, resulting in an overall delay increase of approximately 0.01 s across all network conditions.

Similarly, after integrating a low-pass filter on the slave side, teleoperation master–slave tracking performance underwent continuous, uninterrupted testing for four hours under each network condition. Statistical analysis was conducted on the experimental data of control performance parameters with the introduction of the low-pass filter on the slave side. A global performance analysis of master–slave motion delay and joint motion errors over extended testing periods was performed, and the control performance PDF parameters under different network conditions were plotted, as shown in Fig. 5.9. Comparing the control performance with and without the filter on the slave side, as illustrated in Fig. 5.7, the impact of the filter introduction on master–slave delay slightly increased under near, mid, and far-range Wi-Fi 6 network connections and 5G network connections, as indicated in the left graphs of Fig. 5.9a–d. Particularly for 5G network connections, the introduction of the filter had a more pronounced effect on master–slave delay compared to Wi-Fi 6 network connections. Regarding joint motion peak error parameters, as depicted in the right graphs of Fig. 5.9a–d, it is evident that with the implementation of the low-pass filter on the slave side, the error values under all network conditions were reduced compared to the original unfiltered data input teleoperation. Additionally, the distribution of error

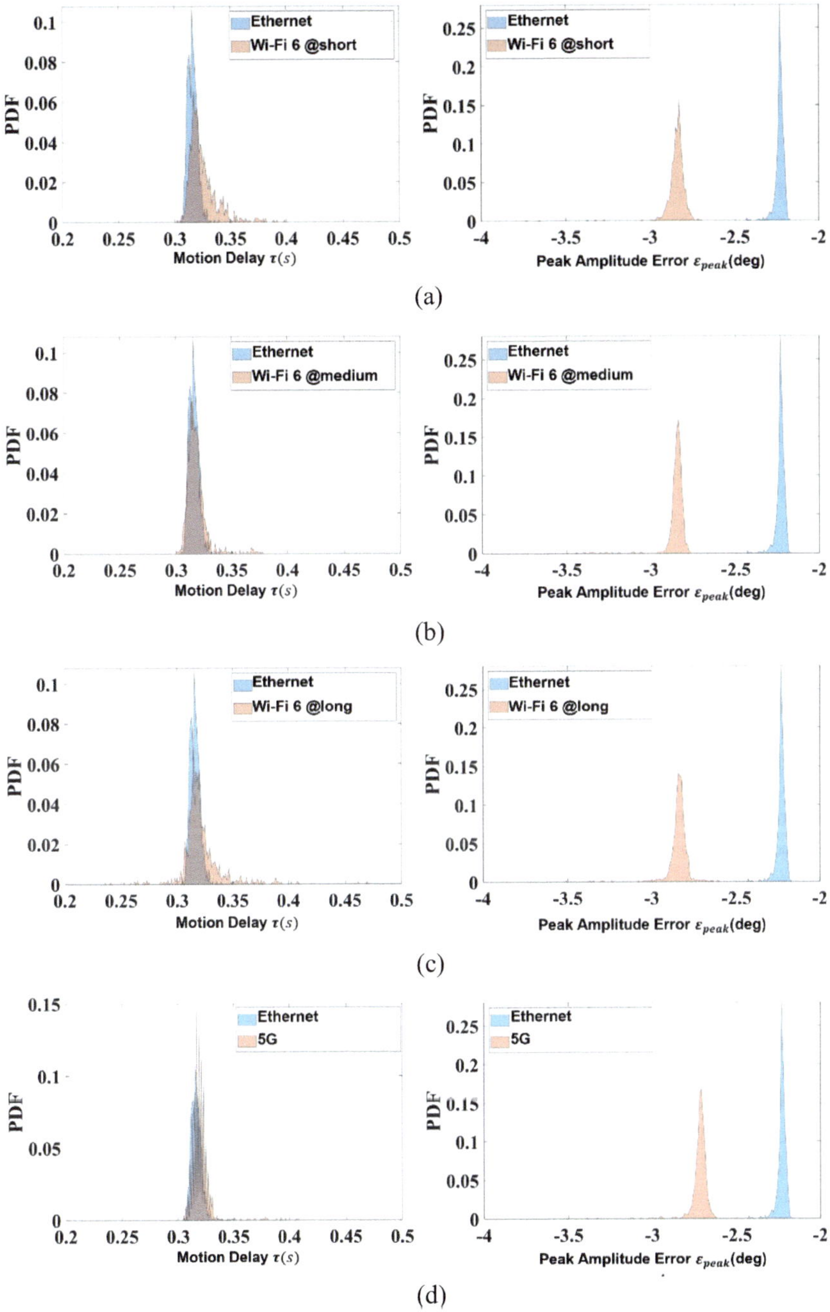

Fig. 5.7 Statistical analysis of robot control using the raw data, reprinted from Ref. [2], CC BY 4.0. **a** PDF curves of control performance under Wi-Fi 6@short network condition. **b** PDF curves of control performance under Wi-Fi 6@medium network condition. **c** PDF curves of control performance under Wi-Fi 6@long network condition. **d** PDF curves of control performance under 5G network condition

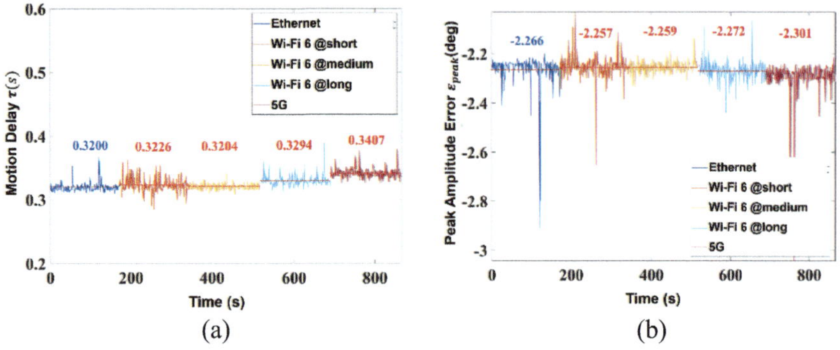

Fig. 5.8 Overview comparison of control performance after deploying a filter, reprinted from Ref. [2], CC BY 4.0. **a** Motion delay. **b** Joint position peak amplitude error

values under the four wireless network connection conditions was lowered to a level nearly consistent with Ethernet.

5.2.3 Communication and Control Co-design for Teleoperation

Building on the performance comparison experimental results, we further computed the Complementary Cumulative Distribution Function (CCDF) curves for master–slave motion delay and peak error during long-term testing of both raw data mapping teleoperation and filtered teleoperation. The comparisons of control performance parameters without and with filters are shown in Figs. 5.10 and 5.11, respectively. The CCDF probability distribution curves for motion delay under four different wireless network conditions reveal that, without filtering, the motion delay distributions across the five network conditions are similar, as illustrated in Fig. 5.10a. However, the CCDF probability distribution curves for the peak error parameter show clear distinctions under different network conditions, as depicted in Fig. 5.10b. The Ethernet wired connection exhibits the smallest concentrated distribution of peak error, while the three Wi-Fi 6 network connections show a larger concentrated distribution of peak error. The 5G network connection has a smaller error distribution compared to Wi-Fi 6, but it is significantly larger than that of the wired connection condition.

The CCDF probability distribution curves for motion delay and peak error after introducing the filter are shown in Fig. 5.11. Compared to the CCDF probability curves of control performance parameters without the filter in Fig. 5.10, it is evident that the introduction of the filter significantly optimizes the probability distribution of the peak error, with the peak error overall tending towards a smaller error distribution. However, at the same time, the distribution of master–slave motion delay slightly

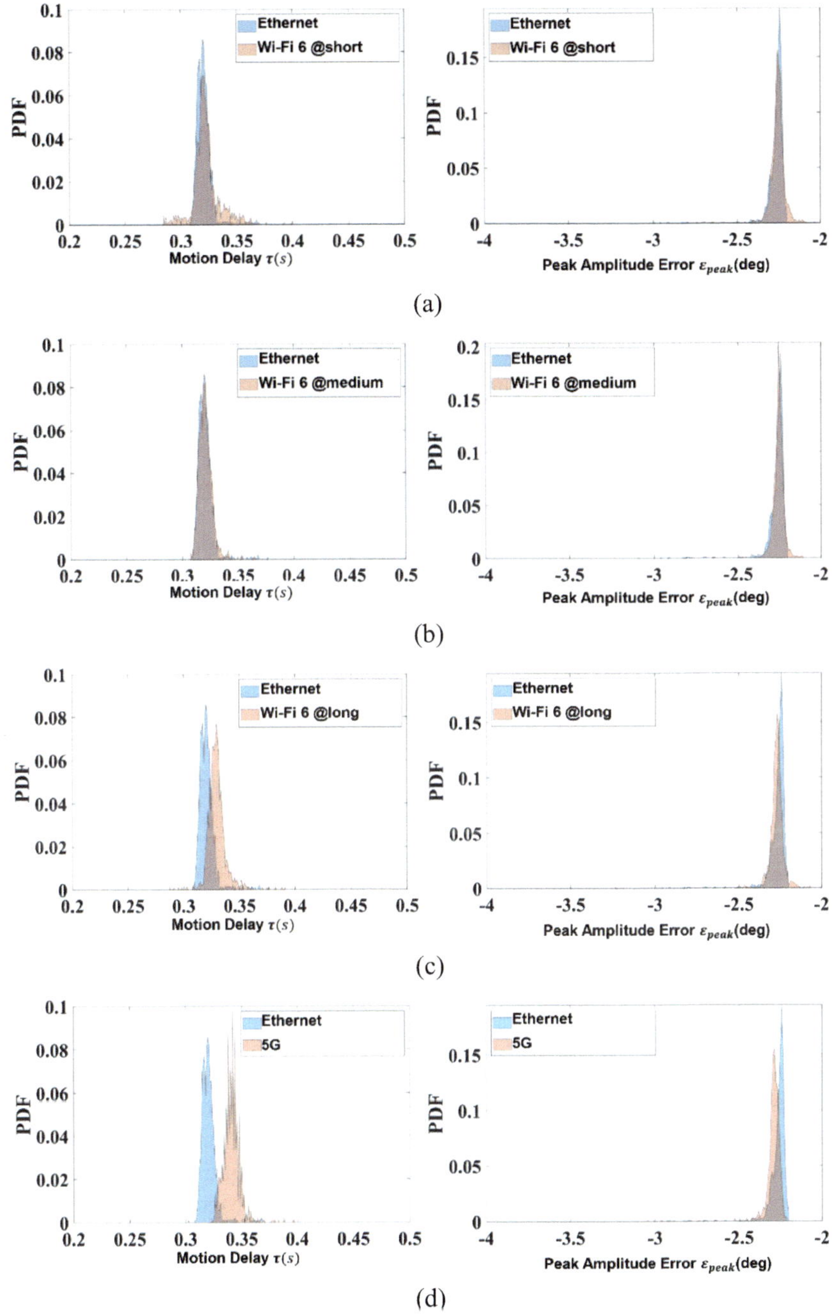

Fig. 5.9 Statistical analysis of control after deploying a filter on robot controller, reprinted from Ref. [2], CC BY 4.0. **a** PDF curves of control performance under Wi-Fi 6@short network condition. **b** PDF curves of control performance under Wi-Fi 6@medium network condition. **c** PDF curves of control performance under Wi-Fi 6@long network condition. **d** PDF curves of control performance under 5G network condition

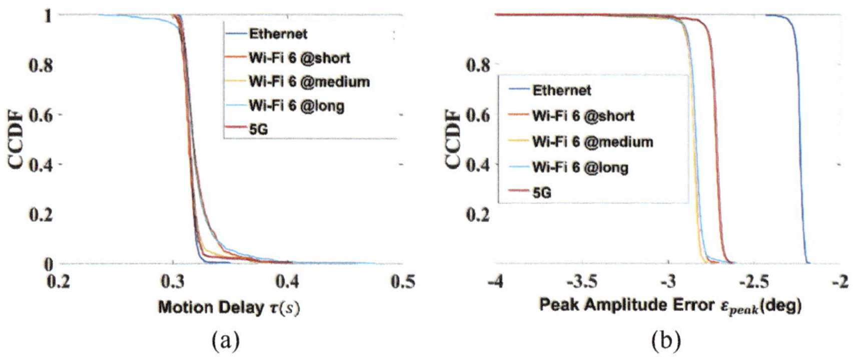

Fig. 5.10 CCDF curves of control metrics without filtering, reprinted from Ref. [2], CC BY 4.0. **a** CCDF curves of motion delay. **b** CCDF curves of joint position peak error

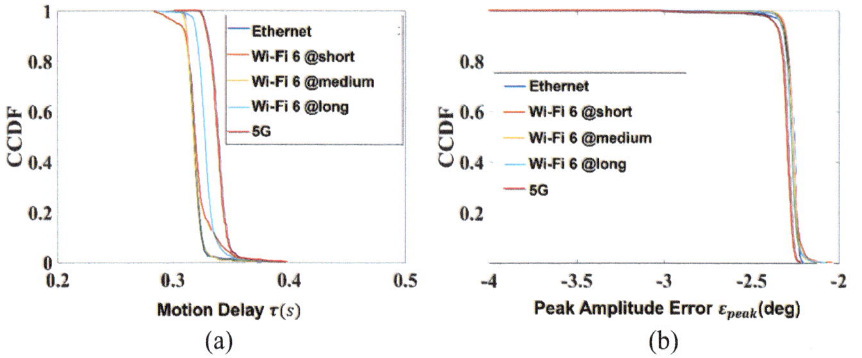

Fig. 5.11 CCDF curves of control metrics with filtering, reprinted from Ref. [2], CC BY 4.0. **a** CCDF curves of motion delay. **b** CCDF curves of joint position peak error

increases, with the most noticeable increase in delay observed under the 5G network condition.

For general control systems, both response speed and accuracy are of paramount concern in practical production scenarios. Consequently, when addressing the needs of teleoperation control systems tailored to diverse application requirements, the selection of an appropriate control strategy to meet specific usage scenarios is summarized as depicted in Table 5.2. In teleoperation systems requiring swift responses, the implementation of end-point filtering results in a slight increase in end-to-end delay. Under such circumstances, the performance gap between LHST and SHLT wireless networks is marginal. However, in scenarios where filtering is not utilized, the delay observed in LHST network conditions is smaller compared to SHLT. Given these performance benchmarks, Wi-Fi 6 is favored over 5G. In teleoperation systems emphasizing precision, the adoption of end-point filtering notably reduces end-to-end response errors. In this context, LHST network optimization surpasses SHLT,

Table 5.2 Network profile selection strategy for different control scenarios, reprinted from Ref. [2], CC BY 4.0

Design expectations	Control strategy	Network profile selection	Examples
Responsiveness	End-point filtering	LHST $\approx$ SHLT	Wi-Fi 6 or 5G
	Without filtering	LHST < SHLT	Wi-Fi 6
Accuracy	End-point filtering	LHST > SHLT	5G
	Without filtering	LHST $\approx$ SHLT	Wi-Fi 6 or 5G

yielding superior error performance. Consequently, 5G emerges as the preferred choice. Nonetheless, the disparity in end-point error performance between LHST and SHLT wireless networks is inconsequential in systems lacking end-point filtering.

5.3 Robot Teleoperation via Wide-Area Wired Networks

5.3.1 The Architecture of the Intercontinental Communication Link

Teleoperation control systems typically use local networks to establish master–slave communication for controlling remote robots. However, stable and reliable intercontinental wide-area network (WAN) communication links are rarely developed. Traditional personal wireless devices or local network cables are insufficient for intercontinental dedicated network connectivity. Instead, dedicated network lines can be established through specific service providers or by using dedicated network servers from cloud providers such as Amazon Web Services (AWS), Microsoft Azure, or Google Cloud Platform (GCP). Compared to local network connections, cloud service-based network frameworks offer efficient resource utilization and simplified management. They also provide a strong foundation for deploying and integrating cloud-based applications and services.

To enable the connection between the control input data from the master operator and the control reception data of the slave robot in a teleoperation system, this study utilizes the virtual private network (VPN) service provided by Google Cloud Platform. As illustrated in Fig. 5.12, an instance with 20 GB of memory and a single-core Intel Broadwell virtual CPU running an Ubuntu system is instantiated on Google Cloud Platform. This instance serves as the server for deploying and hosting the VPN, implemented using OpenVPN [4], which is an application-layer VPN implementation based on the OpenSSL library. Compared to traditional VPNs, it boasts simple and user-friendly configuration. By leveraging the industry-standard SSL/TLS protocol, OpenVPN ensures secure data transmission channels between connected devices.

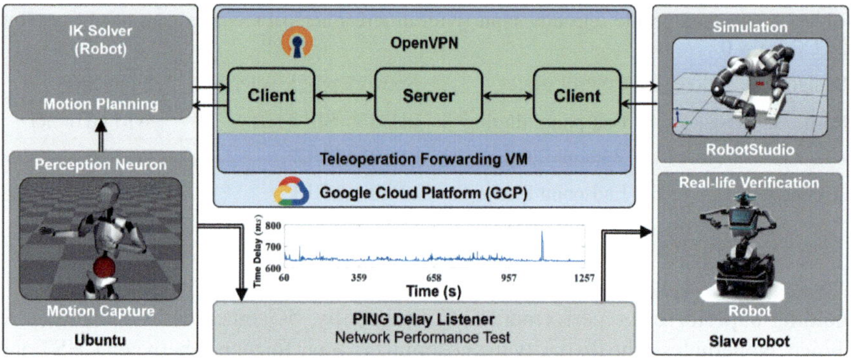

Fig. 5.12 WAN teleoperation framework based on google cloud platform, reprinted from Lyu et al. [1], Copyright (2025), with permission from IEEE

By creating an Ubuntu virtual machine instance on GCP and installing and deploying OpenVPN, you can use the virtual machine as the server side of a VPN. Configuring OpenVPN in point-to-point connection mode enables secure encrypted communication between any client nodes connected to the server. Once the VPN setup is complete, configure the data output terminal of the main operator and the motion data receiving terminal of the secondary robot as VPN clients. The Google Cloud server is set to be located in Frankfurt.

To systematically study the performance of network connections during intercontinental teleoperation, it is necessary to test network loopback latency. However, using a public cloud to build an OpenVPN service does not allow for protocol layer packet analysis. Therefore, a network latency listener based on PING (Packet Internet Groper) [5] needs to be developed, using the IP addresses of two devices on the network route. PING is a service command operating at the application layer of the TCP/IP network architecture, primarily sending ICMP (Internet Control Message Protocol) [6]. Echo request packets to a specific destination host. This enables the testing of end-to-end latency and real-time millisecond-level latency recording, which facilitates subsequent analysis by correlating control performance changes with network performance metrics.

Based on the previously implemented dedicated intercontinental teleoperation data transmission network using a public cloud platform, this section designs an intercontinental long-distance pose mapping teleoperation system. Two test case teleoperation control systems are set up, utilizing the YuMi dual-arm robot and the dual-arm Kinova robotic arm. An intercontinental dedicated network link is established between Västerås, Sweden, and Hangzhou, China, using the Google Cloud Platform and OpenVPN. The operator's motion capture and action data processing are deployed in Sweden as the input for the teleoperation master side, while the controlled robots are deployed in China as the output of the teleoperation system slave side. The straight-line distance between the master and slave sides exceeds 7,800 km.

The intercontinental teleoperation system architecture for the YuMi dual-arm robot is set up as shown in Fig. 5.13. The teleoperation master input side is deployed in Sweden and includes a Windows computer (Intel(R) Core(TM) i7-7500U CPU @ 2.70 GHz × 4 cores) connected to wearable motion capture devices. Additionally, the master side deploys ROS on a NUC10 computer with Ubuntu(Intel(R) Core(TM) i5-10210U CPU @ 1.60 GHz × 4 cores). These computers are connected via a NetGear GS105 Gigabit switch and accessed to the internet. On the slave side, a Windows computer configured with the RobotStudio simulator controls the motion of the YuMi dual-arm robot. This Windows computer is directly connected to the Ethernet. The IP configurations of the master and slave sides are matched to the cloud server node configuration. OpenVPN clients are started on both the master side Ubuntu system and the slave side Windows system to establish the intercontinental network connection between them.

The real-world validation conducted on the dual-arm robot built using the Kinova Jaco robotic arms is depicted in Fig. 5.14. The teleoperation input from the master side, involving human body motion, aligns with the configuration of the YuMi robot teleoperation system. On the slave side, an Ubuntu system NUC10 is deployed with ROS installed to serve as the local robot controller for the Kinova dual robotic arms. The ROS systems on the operator's master side and the robot's slave side are connected to the public cloud virtual private network. They exchange topic data through the shared ROS_MASTER approach, where the ROS system on the robot side serves as the host for the ROS_MASTER, and the ROS system on the master side, processing the operator's motion data, acts as the client.

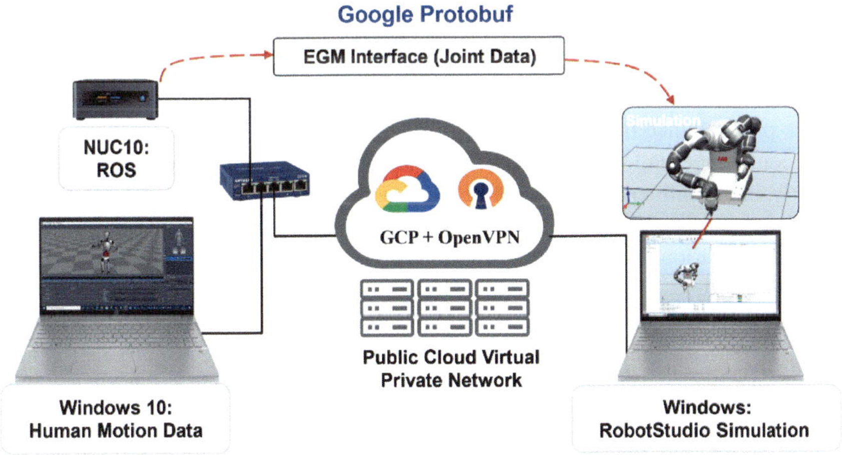

Fig. 5.13 Simulated verification of WAN teleoperation on dual-arm YuMi robot

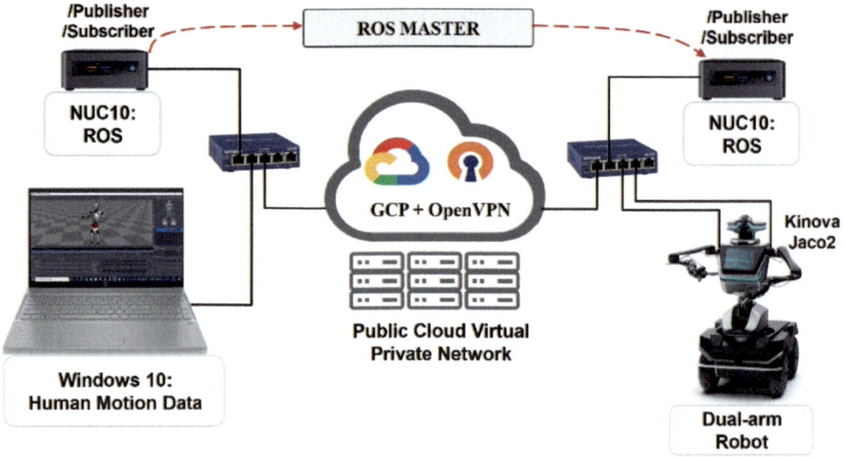

Fig. 5.14 Real-life verification of WAN teleoperation on the anthropomorphic robot

5.3.2 Feedforward Control Framework and the Comparative Results

As depicted in Fig. 5.15, the operator's motion capture and data processing system, along with the controller, are deployed in Västerås, Sweden. Meanwhile, the controllers for the teleoperation robots in both test cases are stationed in Hangzhou, China. The communication between the master and slave sides is facilitated through a VPN established on the Google Cloud Platform. Throughout the testing process, parameters such as motion delay and peak joint position error for both master and slave sides, as well as network latency, are meticulously recorded and analyzed based on the performance indicators outlined in Chapter 5. Experimentally, performance tests commence under local wired connection conditions as a benchmark. Subsequently, intercontinental teleoperation performance comparison tests between China and Sweden are conducted. Each intercontinental experimental test lasts for one hour under various network conditions, and the results are meticulously analyzed and compared.

The experimental results under local Ethernet connection conditions are presented in Fig. 5.16. With minimal network latency in the local setup, the primary factor influencing teleoperation control performance is the latency introduced by the controllers. Analysis of the continuous one-hour testing data reveals an average network latency of 0.316 s for both master and slave sides, with an average peak joint motion error of −2.232°.

As illustrated in Fig. 5.17, compared to the local wired network connection, intercontinental teleoperation experiences a latency exceeding 630 ms. This increase is attributed to the multiple hops involved in establishing the entire virtual private network connection. Additionally, the increased network loopback latency between

Fig. 5.15 Experiment setup and comparison for intercontinental robot teleoperation, reprinted from Lyu et al. [1], Copyright (2025), with permission from IEEE

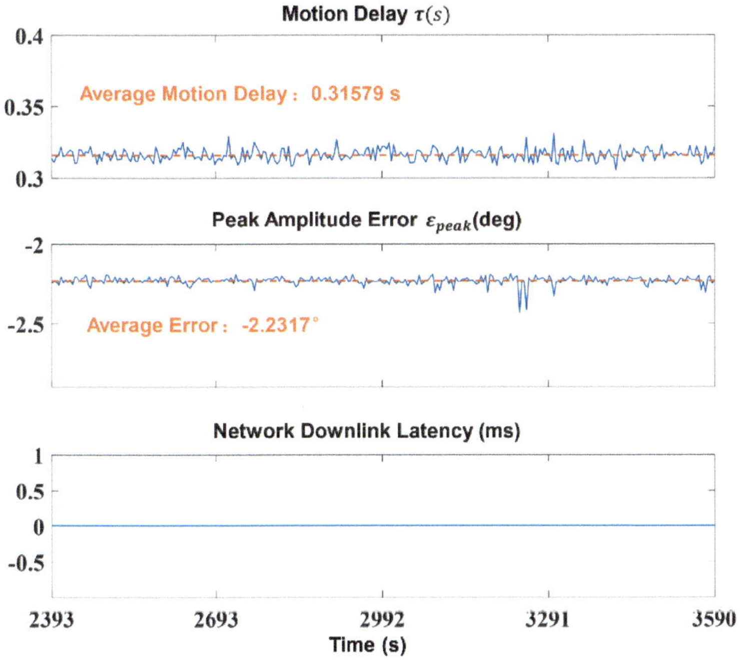

Fig. 5.16 Benchmark performance under local Ethernet connection conditions, reprinted from Lyu et al. [1], Copyright (2025), with permission from IEEE

the master and slave sides leads to a rise in motion delay up to 1.230 s, and the average error value climbs to −2.777°—significantly higher than which under the local wired network connection conditions. Overall, teleoperation control performance deteriorates noticeably under intercontinental network connection conditions, with substantial increases in latency and error values for both master and slave sides.

In the context of large latency teleoperation scenarios across intercontinental long-distance networks, a feedforward controller is devised to further counterbalance errors stemming from network latency. This controller is designed based on a previously measured network latency parameter of approximately 630 ms.

Feedforward control, also known as feedforward regulation, is a widely employed technique in control systems. It involves predicting the discrepancy between the system output and the desired output and making corrections to achieve precise control. Typically, this method introduces a correction signal at the system input to mitigate anticipated errors. The essence of feedforward control lies in introducing a correction signal prior to the system output to offset expected errors. It is most effective in systems that can be accurately modeled and predicted, requiring prior knowledge of system characteristics and anticipated outputs. In the intercontinental teleoperation control system of this study, comprehensive experiments and analyses have

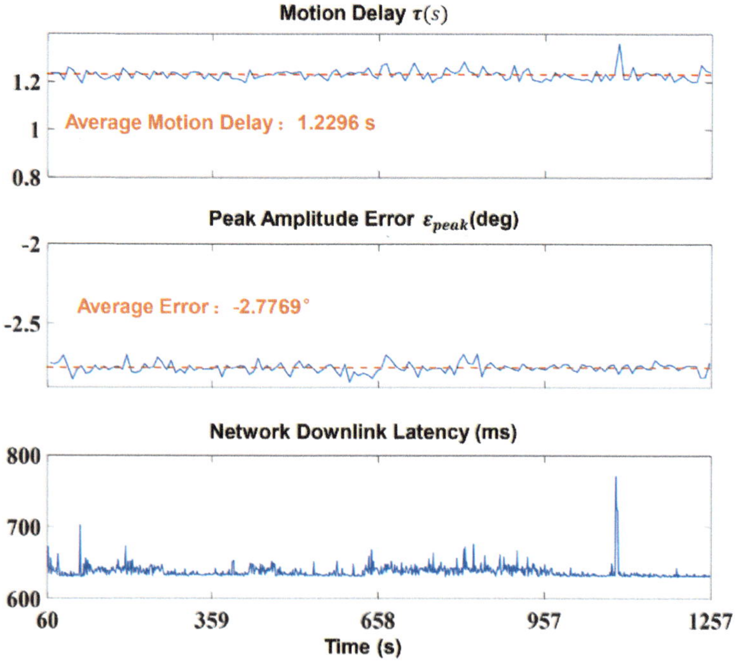

Fig. 5.17 Teleoperation performance under intercontinental WAN conditions, reprinted from Lyu et al. [1], Copyright (2025), with permission from IEEE

scrutinized the network performance parameters and control performance parameters, grounded on the previously measured latency. Hence, integrating a feedforward control strategy into the original master–slave tracking control system appears apt to optimize control performance.

As illustrated in Fig. 5.18, the most fundamental master–slave position mapping teleoperation system can be conceptualized as a simple feedback control system. Real-time feedback of the current actual position of the slave robotic arm compared to the input position from the master side enables closed-loop control. This traditional position feedback tracking control strategy involves interpolating the input position signal from the master side and the feedback position signal from the slave side, and designing a proportional controller to achieve position tracking between master and slave sides. To compensate for the motion delay and error introduced by network latency, a feedforward differential position value is employed to compensate the input feedforward v_{ref} when specifying the slave side position from the master side. Additionally, to mitigate the impact of network fluctuations on the overall control system performance, a low-pass filter is integrated into the existing master–slave closed-loop control loop to enhance system robustness. To simulate and validate the effectiveness of the feedforward controller designed and implemented in this section, the entire system is first modeled using Simulink in MATLAB. Gaussian distribution is used to simulate network fluctuations based on the magnitude of intercontinental network latency measured during the experimental process. Subsequently, comparative experiments are conducted to evaluate teleoperation performance between the original mapping teleoperation system and the system with the introduced feedforward filter design, as depicted in Fig. 5.19.

The curves depicted in Fig. 5.19a represent the experimental results of the original feedback control loop in the simulated system, illustrating the master–slave mapping. In the simulation, the network latency is denoted as $\text{avg}(RTT) + |X_t|$,

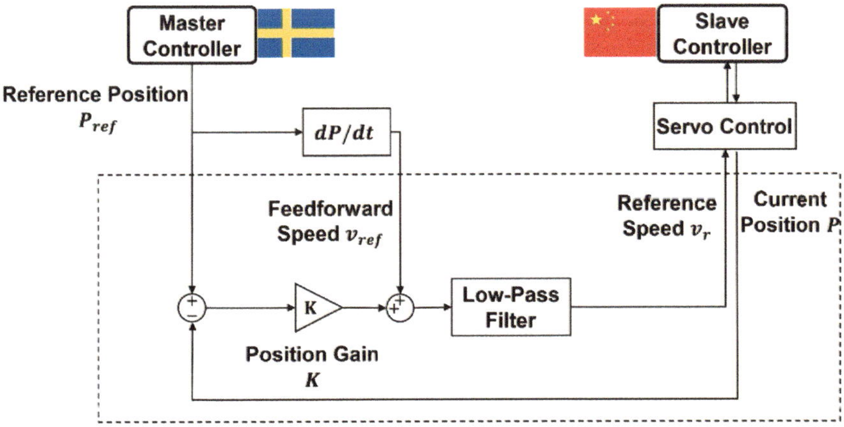

Fig. 5.18 Feedforward controller for WAN teleoperation with high latency, reprinted from Lyu et al. [1], Copyright (2025), with permission from IEEE

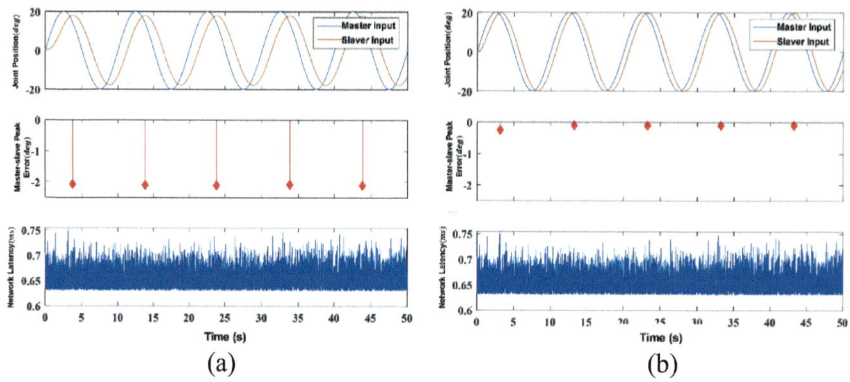

Fig. 5.19 Simulation comparison for feedforward control framework. **a** Initial system simulation. **b** Simulation with feedforward controller

where avg(RTT) represents the mean delay of the intercontinental network loop obtained during preliminary testing, and "t" is modeled based on the fluctuation distribution of network latency, using a Gaussian distribution. From the comparative simulation experiments conducted using Simulink, it is evident that due to the influence of the control loop's loopback delay, which includes both network latency and control latency, the slave-side input noticeably lags behind the master-side position input. Additionally, an analysis of the errors between each peak value for each control cycle reveals an approximate error level of 2° for both master and slave sides, as depicted in the middle plot of Fig. 5.19a. Introducing the feedforward controller designed in this section yields significant improvements. Comparing the position signals between master and slave sides, it can be observed that, relative to the control effect without feedforward control shown in Fig. 5.19a, the motion delay between master and slave sides is reduced. Moreover, the error values are further diminished, as demonstrated in the middle plot of Fig. 5.19b, indicating a substantial decrease in peak error values.

When further integrating the designed feedforward controller into the actual intercontinental teleoperation control system, the test results are shown in Fig. 5.20. Compared to the performance of the intercontinental network teleoperation loop with the initial proportional controller depicted in Fig. 5.17, it is evident that the introduction of the feedforward controller significantly improves the system. During the actual intercontinental teleoperation control tests with the feedforward controller, the average motion delay is 1.282 s, which is comparable to the original teleoperation system's performance. Additionally, the average peak error value is −2.199°, representing a significant reduction in error compared to the teleoperation control system without the feedforward controller.

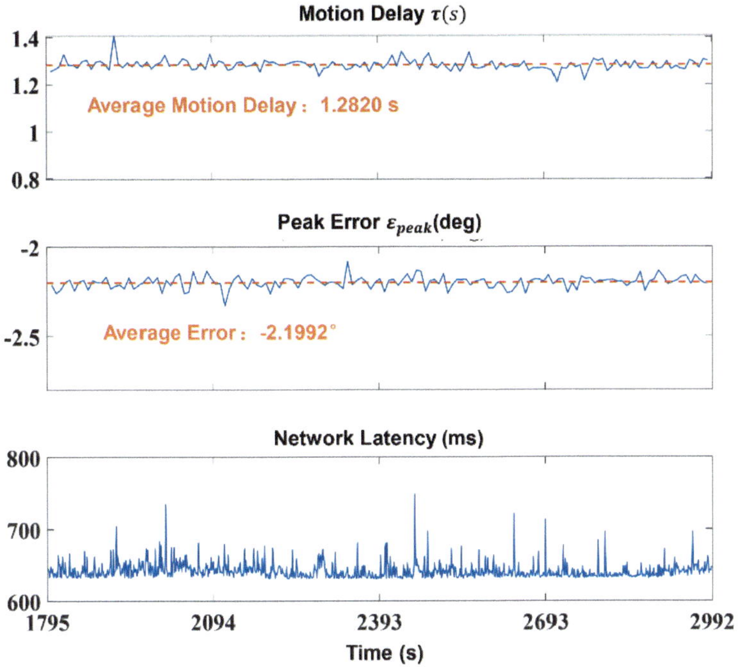

Fig. 5.20 Real-life performance of intercontinental teleoperation with feedforward, reprinted from Lyu et al. [1], Copyright (2025), with permission from IEEE

5.3.3 Verification of an Intercontinental Teleoperation System

To further validate the practical application performance of the intercontinental teleoperation control system, a bimanual coordinated insertion task was conducted using the lab-constructed dual-arm robot teleoperation control system. As illustrated in Fig. 5.21, the operator in Sweden wears motion capture equipment that records upper limb movement data and performs kinematic calculations. The control data output is managed using the incremental pose mapping teleoperation technique designed in Chap. 4. The operator has two visual feedback screens: the main screen provides feedback from the primary operation perspective of the robot, allowing the operator to monitor the robot's overall movement, while the second screen shows the camera view of the robotic arm's end effector, giving the operator a clear view of the robot and its end effector's motion control status. The dual-arm robot is located in a laboratory in Hangzhou, China. A camera is positioned to provide a first-person perspective for the robot, and a RealSense RGB-D camera is mounted on the end effector of the Jaco2 robotic arm to capture and feedback the operational view of the end effector. In front of the robotic arm, there is an operation table for placing the objects to be manipulated.

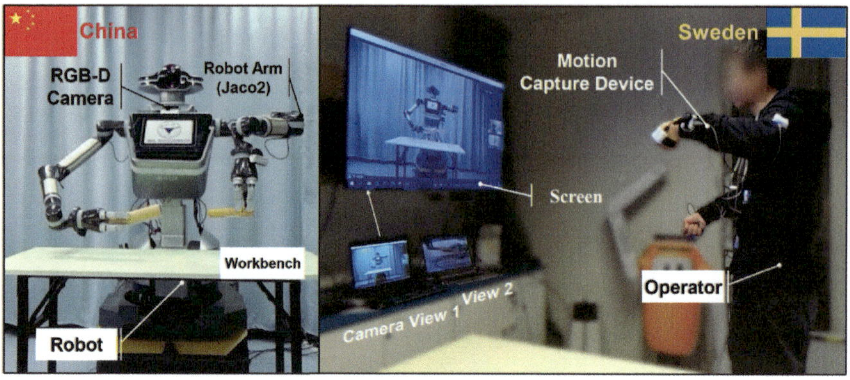

Fig. 5.21 Intercontinental teleoperation system setup and the demonstration, reprinted from Lyu et al. [1], Copyright (2025), with permission from IEEE

In this subsection, the bimanual coordinated insertion task involves manipulating a cylindrical shaft and a cylindrical sleeve. The objective is to insert the cylindrical shaft into the cylindrical sleeve. This task is designed to be a loose fit, with the primary goal of validating the position control performance of the intercontinental teleoperation control system, without using end-effector force control for adjustments. As shown in Fig. 5.22, in the initial state, the right arm of the robot grips the cylindrical sleeve, while the left arm is in its initial position. The operator adjusts the left arm's posture to align with the workpiece on the table in preparation for grasping. The operator then moves the left hand downward to control the left arm's end-effector to the gripping position of the cylindrical shaft on the table. Using a gesture to enable the end-effector, the operator grips the cylindrical shaft and lifts it. At the 10-s mark, the operator coordinates the movements of both arms to fine-tune the positions of the left arm holding the cylindrical shaft and the right arm holding the cylindrical sleeve. Based on feedback from the two camera perspectives, the operator adjusts the posture and position of the cylindrical shaft to align it precisely with the hole of the cylindrical sleeve. The operator then slowly moves and continuously adjusts the posture of the left arm's cylindrical shaft to insert it into the sleeve. Finally, the operator releases the grip of the end-effector, completing the coordinated insertion task, and resets the robotic arms.

Additionally, to further demonstrate the practicality of the designed teleoperation system, functional tests were conducted in various household operation scenarios within an indoor environment, as shown in Fig. 5.23. The figure illustrates the operator in Sweden remotely controlling the robotic arm in China to perform tasks such as grasping and placing wooden blocks into corresponding boxes, organizing and sorting fruits on a table, and using a wireless Bluetooth stethoscope to perform cardiac and pulmonary auscultation on a remote patient.

To further validate the practicality of the designed intercontinental teleoperation control system, a master-side feedback and safety interaction system was implemented using capacitive sensing robotic skin. This system underwent functional

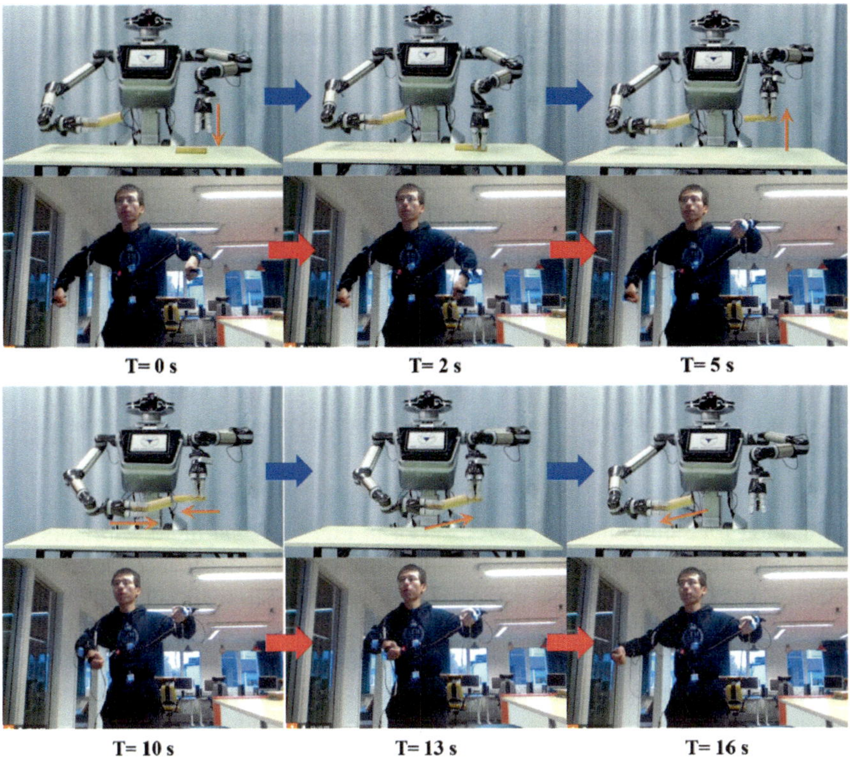

Fig. 5.22 Intercontinental teleoperation for bimanual coordinated insertion task, reprinted from Lyu et al. [1], Copyright (2025), with permission from IEEE

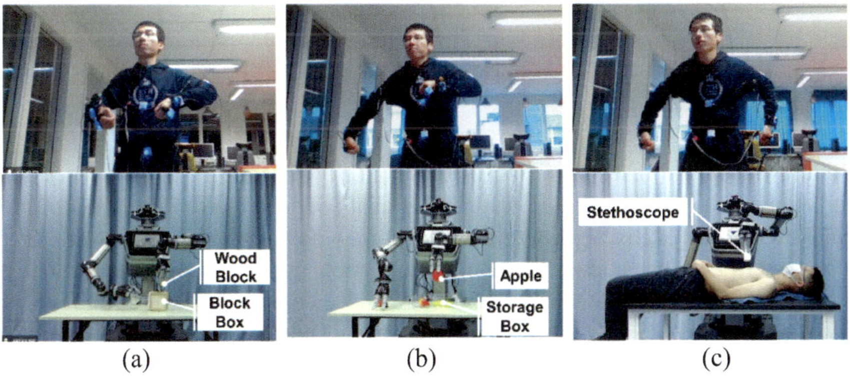

Fig. 5.23 Some use cases of the intercontinental teleoperation. **a** Teleoperated wooden block organization, **b** Teleoperated fruit handling, **c** Teleoperated auscultation verification

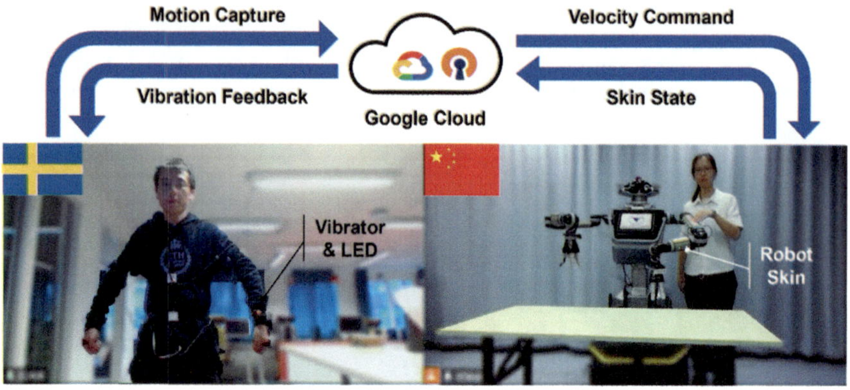

Fig. 5.24 Intercontinental sensing feedback and safety interaction for teleoperation, reprinted from Lyu et al. [1], Copyright (2025), with permission from IEEE

verification for safety interaction. As shown in Fig. 5.24, flat motor vibrators and LEDs were attached to the operator's wearable device in Sweden to indicate vibration status. The robot in China was equipped with sensory skin to detect the proximity of objects or the operator. The sensory data was transmitted through the Google Cloud intercontinental teleoperation network to the operator's side in Sweden. Upon receiving the feedback signal, the processor on the operator's side activated the vibrators and LEDs. During remote operation, the operator controls the robot arm using motion capture, sending velocity commands to the remote arm to perform various actions and tasks. This feedback control mechanism enables safe interaction between the robot arm and its environment, enhancing the operator's perception of the remote operation site.

5.4 Summary

Building upon the research on pose-mapping teleoperation techniques discussed earlier, this chapter focuses on testing and analyzing the performance of teleoperation control systems under wireless local area network (LAN) and wide area wired network connections. For the teleoperation systems using local wireless network connections, statistical analysis is conducted on real-time sniffed uplink and downlink latencies to characterize the performance under various Wi-Fi 6 and 5G conditions. Additionally, different control-demand scenarios are studied to further investigate the collaborative design methods for network and control in teleoperation systems, providing data support for teleoperation design in various functional scenarios. While for the teleoperation systems using wide area wired network connections, tests are conducted to characterize the performance of intercontinental wide area network connections and real-time control. An optimization method using feedforward control for ultra-long-distance teleoperation control systems is proposed. This facilitates teleoperation across continents, such as from Sweden to China, under intercontinental wide area network connections. The practical performance of the intercontinental teleoperation control system is validated through multiple remote operation tasks. The main conclusions of this chapter are as follows:

(1) A semi-physical simulation testing system has been devised to assess the performance of local wireless network control. This system integrates real network deployment environments into the teleoperation testing setup. Extensive tests were conducted on teleoperation control under various conditions, including Wi-Fi 6 networks at close, medium, and far distances, as well as 5G network connections. Analysis and statistical evaluations have been carried out on network latency in both uplink and downlink directions, as well as on the motion delay and joint peak error of the control system's master–slave ends. The findings reveal that Wi-Fi 6 networks exhibit a higher probability distribution of latency exceeding 10 ms compared to 5G networks, although the average latency of Wi-Fi 6 networks is lower. Regarding control performance, the latency of master–slave ends under wireless local area network connections is notably higher than that under wired connections, accompanied by larger joint peak errors. Furthermore, the joint peak error of master–slave ends under Wi-Fi 6 wireless network connections surpasses that under 5G network connections.

(2) To tackle the issue of decreased control performance stemming from network communication delays in teleoperation systems, a low-pass filter has been implemented at the slave end of the system. This filter aims to mitigate the impact of network transmission uncertainties on control performance metrics. Experimental findings reveal that introducing a filter at the slave end notably reduces peak joint motion errors under wireless local area network conditions. However, it simultaneously leads to an increase in motion delay between the master and slave ends. Additionally, we further outline co-design approaches for teleoperation systems and networks tailored to different requirements: for teleoperation systems requiring swift response, omitting a filter at the slave end is advisable, and networks like Wi-Fi 6, characterized by low-latency and high-throughput capabilities, are preferred; for teleoperation systems requiring high precision, introducing a filter at the slave end is necessary, and networks like 5G, suitable for precision control, should be chosen.

(3) A dedicated network connection for ultra-long-distance teleoperation spanning 7,800 km from Sweden to China has been meticulously established. Extensive testing and comparative analysis are conducted to assess the performance of the intercontinental ultra-long-distance teleoperation control system and its network connectivity. The real-time impact of network performance on robot teleoperation systems across ultra-long-distance wide area network connections is thoroughly examined. Results revealed an average end-to-end delay of 1.230 s for the ultra-long-distance wide area network connection, with an average peak error of -2.777°. Additionally, with a master velocity feedforward

control method which leveraging real-time position feedback from the slave-end robot the average peak error between the master and slave ends is effectively reduced to $-2.199°$, marking a significant 20.80% reduction compared to error levels in loops without feedforward control. Furthermore, diverse application-level tasks and functionalities are meticulously designed and implemented for various teleoperation interaction scenarios. The implementation allows operators wearing motion capture devices in a lab in Sweden to seamlessly control a dual-arm robot situated in a lab in China. Overall, these validations underscore the practicality and effectiveness of the intercontinental teleoperation control system developed in this chapter.

References

1. Lyu H, Zhou H, Wang R et al (2023) Towards intercontinental teleoperation: a cloud-based framework for ultra-remote human-robot dual-arm motion mapping. In: Proceedings of the intelligent robotics and applications, Hangzhou, China, pp 132–144
2. Lyu H, Pang Z, Bhimavarapu K et al (2023) Impacts of wireless on robot control: the network hardware-in-the-loop simulation framework and real-life comparisons. IEEE Trans Industr Inf 19(9):9255–9265
3. Smołka I, Stój J (2022) Utilization of SDN TECHNOLOGY FOR flexible EtherCAT networks applications. Sensors 22(5):1944
4. Skendzic A, Kovacic B (2017) Open source system OpenVPN in a function of virtual private network. In: Proceedings of the IOP conference series: materials science and engineering, San Francisco, USA, p 012065
5. Jackson G, Keleher P, Sussman A (2015) A ping too far: real world network latency measurement. In: Proceedings of the IEEE international conference on e-science, Munich, Germany, pp 580–588
6. Donta PK, Srirama SN, Amgoth T et al (2022) Survey on recent advances in IoT application layer protocols and machine learning scope for research directions. Digit Commun Netw 8(5):727–744

Chapter 6
Healthcare Applications of Human-Motion Based Robot Teleoperation

Abstract With the rapid aging of population and the increasing number of disabled elderly individuals, there is a growing demand for elderly care services, creating a significant market for service-oriented robots. This study investigates the use of robot pose-mapping teleoperation technology in remote homecare and telemedicine applications. A homecare robot prototype, featuring a dual-arm collaborative robot (YuMi) and an omnidirectional mobile chassis, is designed to assist elderly individuals with tasks such as retrieving medicine and delivering items within the home. The robot's teleoperation system is based on operator motion capture, enabling intuitive control and efficient task execution. Additionally, the study explores the application of this technology in healthcare, particularly during the COVID-19 pandemic, by demonstrating its use in isolation wards to reduce healthcare workers' exposure to patients. The teleoperated robot can perform essential tasks such as food delivery, disinfection, and remote consultations with healthcare professionals. The findings show that the proposed robot system not only enhances efficiency in elderly care but also offers a practical solution to reduce the risks associated with direct contact between healthcare workers and patients. The research highlights the potential of pose-mapping teleoperation to revolutionize homecare and telemedicine, improving both safety and operational efficiency.

Keywords Teleoperation use cases · Homecare robot · Medical assistive robot · Elderly care · COVID-19

6.1 Teleoperation Use Case for Remote Homecare

With the increasingly serious aging of China's population, the number of disabled elderly is increasing, as well as the burden of social welfare. The demand for social security and services is becoming more and more urgent, which provides a huge market and application scenario for service-oriented robots [1]. It is estimated that by 2030, the number of people aged 60 and above will soar to 329 million, while the number of disabled elderly in China will increase to 67.41 million [2]. While the

H. Lyu et al., *Human Motion Awareness and Robot Teleoperation*, Advanced Topics in Science and Technology in China 4, https://doi.org/10.1007/978-981-96-6545-7_6

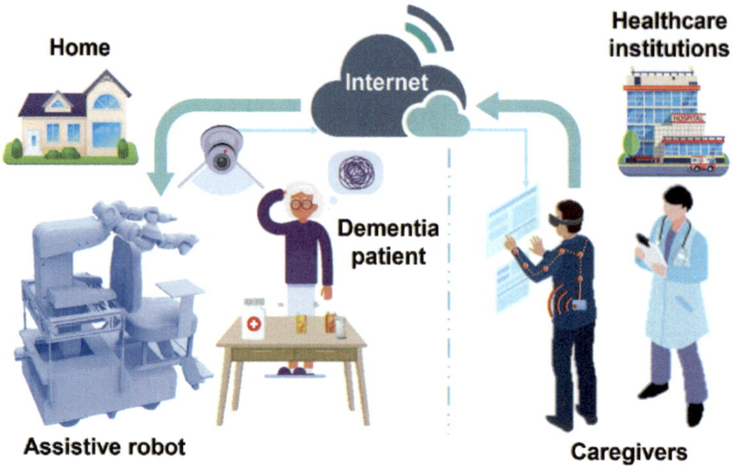

Fig. 6.1 Architecture of the telerobotic system for remote homecare, reprinted from Ref. [1], CC BY 4.0

absolute number of young and middle-aged people is decreasing, the relative number of working people is also decreasing, making the gap in the supply of caregivers in the nursing market larger. Nursing work itself has the characteristics of high labor intensity but low wage level. Therefore, professional elderly care workers have been difficult to hire. Besides, the mobility of care workers is large, which is easy to lose, resulting in a huge market demand for homecare robots [3]. As shown in Fig. 6.1, the development of a new generation of network technology and remote operation technology provides new application scenarios and ideas for care assistance robots, this book proposed a remote homecare application architecture to make up for the shortage of elderly caregivers. Professional caregivers in hospitals, community clinics, and other medical institutions wore remote operation equipment to make corresponding auxiliary actions, controlling the elderly assistive robot located in the home environment through the remote network connection. The disabled elderly in the home environment can complete the operation of taking medicine and moving indoors with the support of remote teleoperation assistance, so as to improve the ability of self-care.

6.1.1 System Architecture of a Homecare Robot Prototype

In this book, an assistive robot prototype for remote homecare was integrated and designed. The overall structure of the prototype is schematically shown in Fig. 6.2, which includes a robotic arm, a mobile chassis, a lifting mechanism, and a ride section. A YuMi dual-arm collaborative robot is used as the actuator of the robotic arm. The omnidirectional mobile chassis is used to carry the rider, the upper part of

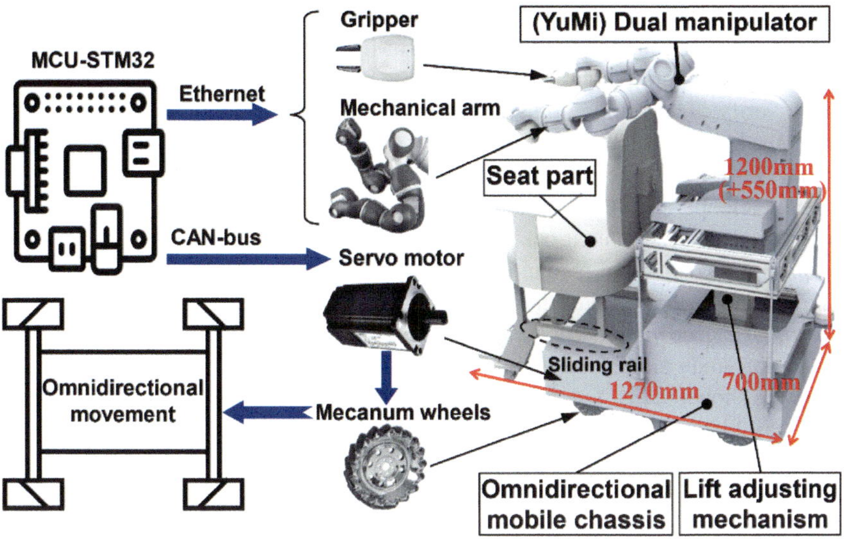

Fig. 6.2 Prototype design of the homecare robot, reprinted from Ref. [4], CC BY 4.0

the YuMi dual-arm robot, and to support the lifting mechanism. The mobile chassis selects the McNamee wheel multi-wheel drive structure to realize omnidirectional movement, which can meet the demand for flexible movement and steering in the narrow space of the home environment. There is a lifting mechanism between the robot arm and the mobile chassis, which can adjust the relative height of the robot arm and the ride section, and the adjustable stroke is 550 mm. The ride section is located in front of the robot arm, and there is a slide mechanism between the arm and the mobile chassis, which can adjust the relative distance between the arm and the robot arm to adapt to the different user body sizes and to adjust the operating space of the robot arm. The robotic arm is installed at the rear of the ride section, and the robotic arm makes corresponding nursing actions from behind the rider. On the one hand, this arrangement makes the overall structure of the nursing robot compact and effectively improves the power density of the service robot. On the other hand, the double robot arm can also make the nursing robot realize more complex nursing actions.

6.1.2 Application for Remote Dementia Care in Home Environments

In order to verify the practicality of the pose mapping teleoperation system in homecare scenarios, functional verification was carried out by teleoperating the YuMi robotic arm to perform item grasping experiments in a laboratory environment. The operator wore the inertial motion capture device PN, made the corresponding upper limb grasping action, and controlled the YuMi robot to deftly grasp the medicine bottle. The teleoperation grasping verification process is shown in Fig. 6.3. The test was to verify the grasping ability of common items in home application scenarios during the bitmap teleoperation process. The operator and the robot were in the same test site without remote feedback equipment. The test result showed that the pose mapping teleoperation technology proposed in this book can accurately transfer the operator's actions and intentions from the master to the slave robot, and successfully complete the grasping of medicines on the desktop.

In order to further validate the practical application capability of the robotic system for care assistance in remote smart homecare scenarios, this book carried out a system application study on the remote home assistance delivery task for a disabled elderly person, as shown in Fig. 6.4. The remote healthcare operator wore a wearable device to capture his own movement data and connected remotely to the local controller of the care-assisting robot in another room via the network. The care recipient is an elderly disabled person riding on the care-assisting robot in the other room. The operator and the care recipient were separated from each other by a distance of 50 m. A webcam was set up at the operation site of the robot to transmit real-time live image data to the operator. The operator adjusted his own movements in real-time by the on-site feedback displayed on the screen, successfully helping the elderly to take the daily items such as water cups on the table and operating the robot arm

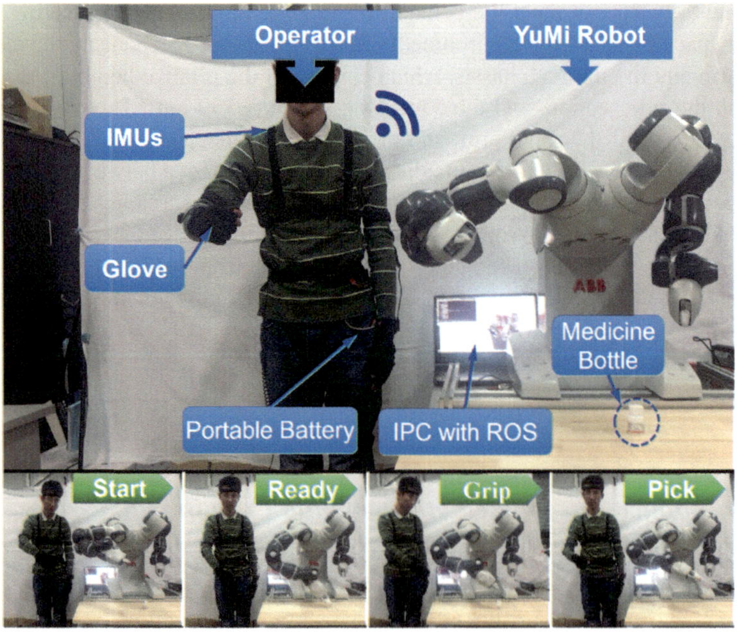

Fig. 6.3 Demonstration of picking up a medicine bottle remotely, reprinted from Ref. [1], CC BY 4.0

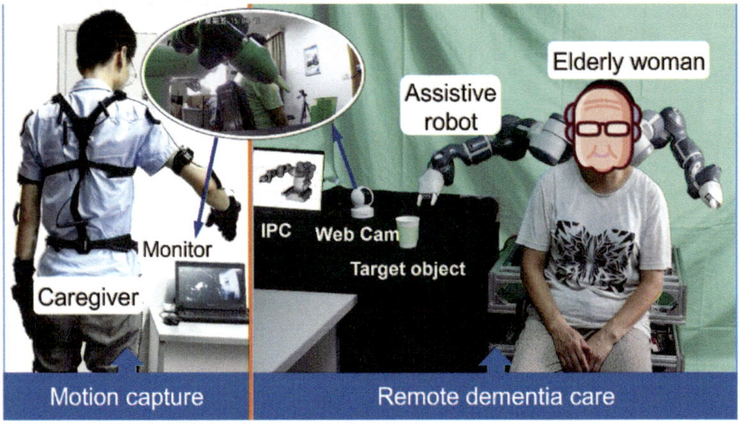

Fig. 6.4 Delivering items remotely for the elderly by robot teleoperation, reprinted from Ref. [1], CC BY 4.0

to deliver them to the elderly who are inconvenient to move. Through the practical application verification, the interface friendliness and operation convenience of the pose mapping teleoperation technology proposed in this book for the main operator user were verified, and the task execution capability of the remote homecare robot was also verified, which proved the application prospect of the remote homecare application framework supported by the posture mapping teleoperation technology proposed in this book.

6.2 Teleoperation Use Case for Telemedicine in COVID-19

During the COVID-19 epidemic, there was a risk of nosocomial infection among healthcare workers in the isolation wards, and many cases of nosocomial infection among healthcare workers due to contact with patients in China. Therefore, the major challenge is to reduce the contact of healthcare workers with patients in the process of diagnosis and treatment, so as to avoid nosocomial infections [5]. The application of remote operation technology can help healthcare workers carry out medical assistance remotely, effectively protecting the life safety of healthcare workers. However, with the current mainstream teleoperation solution, it is still an urgent problem to improve the convenience and operability of the robotic teleoperation control process and reduce the learning cost of healthcare workers to facilitate their quick start [6]. In this section, robot pose mapping teleoperation technology was applied to the first-line diagnosis and treatment of the novel coronavirus, aiming to reduce the risk of infection caused by the frequent entry of medical staff into the isolation ward, reducing the workload and burden of medical care, improving the efficiency of diagnosis and treatment. The proposed telemedicine nursing framework for isolation wards is

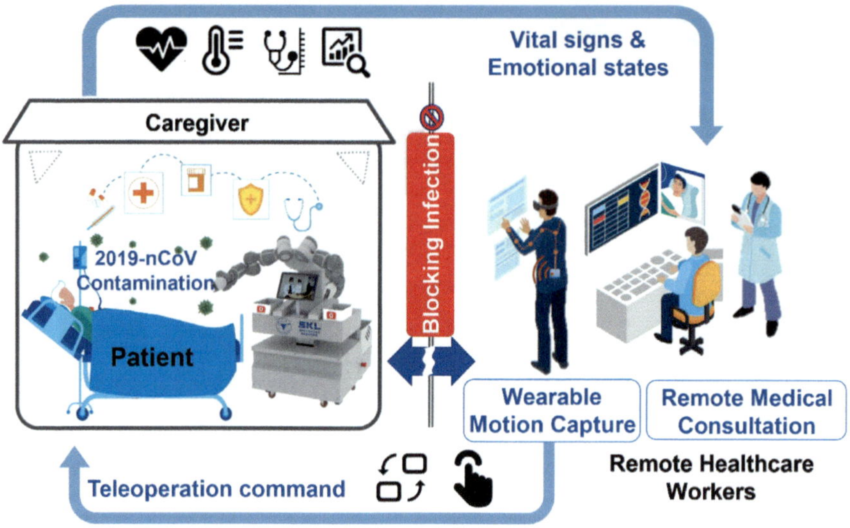

Fig. 6.5 Teleoperated robot in isolation ward for COVID-19 prevention and control, reprinted from Ref. [7], CC BY 4.0

shown in Fig. 6.5. Healthcare personnel can remotely operate the medical assistant robot outside the isolation ward to assist medical personnel in completing part of the ward care work, or the robot can independently complete the daily care work of the patient, blocking the cross-infection caused by the contact between healthcare personnel and the patient. Through real-time monitoring, recording and controlling patients' physical signs and parameters, traceable historical data are provided for later interventional treatment and promoting patients' health recovery, providing a strong guarantee and key technical support for fighting the epidemic.

6.2.1 A Medical Assistive Robot Prototype Used in Isolation Ward

The medical assistance robot in the isolation ward built based on the dual-arm robot YuMi has a human-like two-handed dual-arm structure, as shown in Fig. 6.6. The structure is mainly divided into four parts: an omnidirectional mobile chassis, a two-armed collaborative robot on top of the chassis, a height-adjustment mechanism, and other auxiliary devices. The chassis allows omnidirectional movement through different combinations of movements of the four McNamee wheels, which is suitable for flexible movement in the confined space of an isolation ward. The two-armed robot YuMi is connected to the mobile chassis by a motorized height-adjustment mechanism, which is used to adjust the height of the robot arm and its workspace in unstructured operating scenarios. The design of the omnidirectional

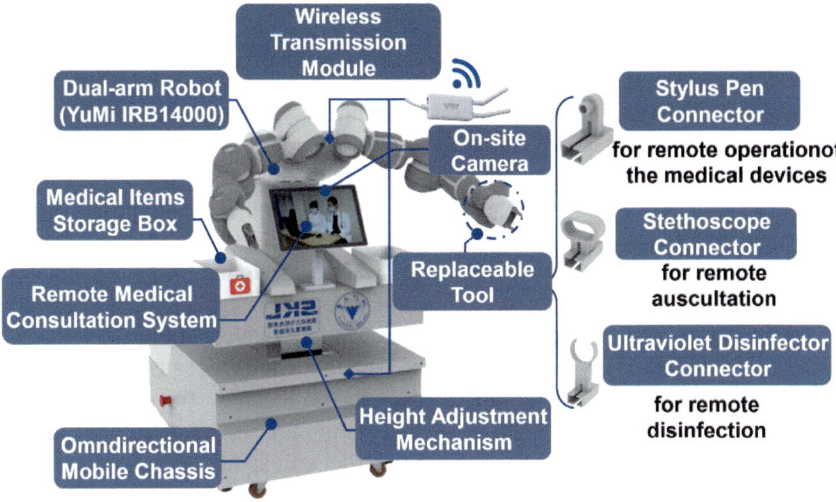

Fig. 6.6 Teleoperated robot prototype for medical assistance in isolation ward, reprinted from Ref. [7], CC BY 4.0

mobile chassis and height-adjustment mechanism greatly extends the movement flexibility of medically assistive robots in isolation wards and hospitals.

In order to meet the daily patient care needs, the end replaceable clamps of commonly used tools (capacitive stylus, Doppler ultrasound equipment, handheld sterilization equipment, etc.) are designed. Based on the smart tablet PC mounted on the front of the robot and equipped with a remote consultation system, the patient can have video interaction and non-contact consultation with the remote healthcare personnel. At the same time, the design of the end-replaceable fixtures allows for the installation of other small diagnostic equipment according to operational needs, assisting healthcare workers in completing tasks such as food and drug delivery, daily decontamination, simple instrumentation, and remote cardiopulmonary auscultation, which can effectively minimize the risk of healthcare workers being exposed to the novel coronavirus. A storage box for medicines, disinfectants, and other essentials is mounted on both sides of the robot. Remote data transmission is established through a pair of high-power Wi-Fi repeaters between the robot and the healthcare worker, allowing for a reliable wireless connection between the healthcare operator and the robot over a distance of 300 m, meeting the needs of telemedicine assistance in the hospital environment.

6.2.2 *Clinical Trials of Robot Teleoperation in Isolation Ward*

During the epidemic period, the medical assistance robot platform designed in this book was tested in the teaching and research Center of the First Affiliated Hospital of Zhejiang University Medical School and tested in the emergency center of the First Affiliated Hospital of Zhejiang University Medical School on the premise of obtaining the approval of clinical experiment ethics review. Figure 6.7 shows the working scenario of the remote operation of medical assistance robot for remote drug delivery. The medical staff wore an inertial motion capture device in a secure work area outside the isolation ward, and the dual robotic arms of the medical assistance robot were remotely controlled by the medical staff through the proposed GuLiM hybrid intuitive mapping method. The camera installed on the robot transmitted the scene vision to the display device at the scene where the nursing worker was located, so as to realize the infection intervention between the medical operator and the infected patient. The purpose of this application is to provide a remote solution that allows healthcare workers to avoid entering isolation wards when treating patients, thereby reducing the risk of cross-infection among healthcare workers in the hospital. On the one hand, in the field verification application process of isolation ward, the YuMi cooperative robot used can monitor external collisions through joint current estimation torque, which has excellent safety performance and collision detection and control mechanism, so it can ensure immediate stop when accidental collisions occur. On the other hand, the proximity sensing robot skin based on the principle of self-capacitance can be applied to the surface of the robot body [8]. The proximity of the human body is perceived before the collision, and the corresponding obstacle avoidance action is made in combination with the manipulator controller to improve the safety of the interaction between robots and people.

In the actual clinical application process, the specific functions of the teleoperation medical assistance robot system proposed in this book are as follows:

(1) Teleoperated item delivery: The movement of the robotic arm is controlled by capturing the movement of the operator's upper limbs. The opening and closing of the end clamp is controlled by capturing the movement of the operator's fingers. Necessary items (such as medicines, thermometers, disinfectants, etc.)

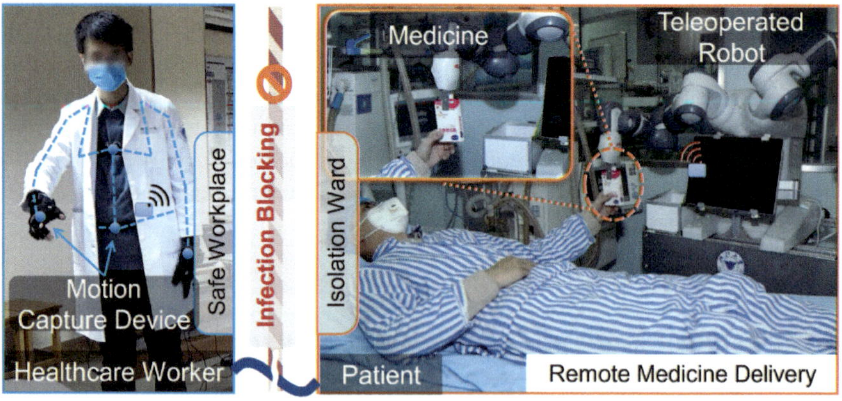

Fig. 6.7 Remote medicine delivery using the teleoperated robot

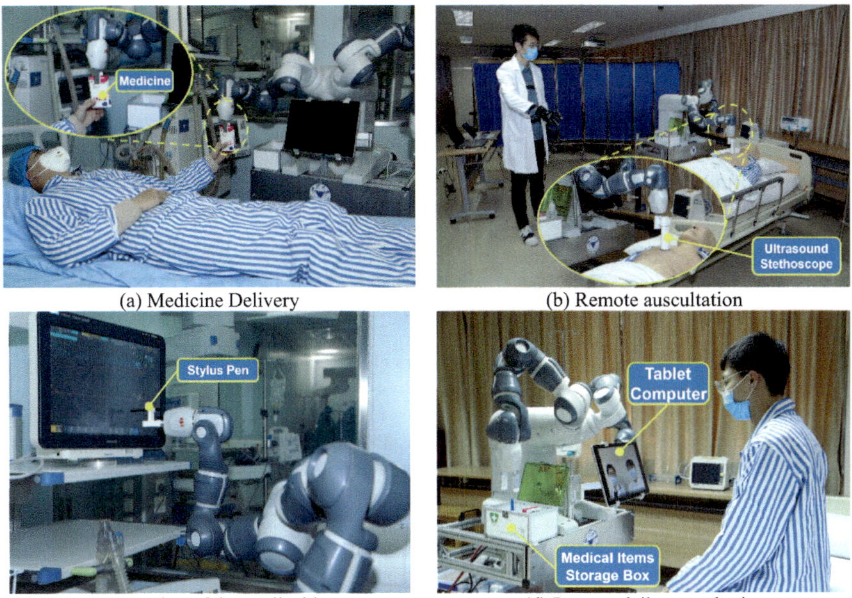

(a) Medicine Delivery (b) Remote auscultation (c) Remote operation of the medical instruments (d) Remote daily consultation

Fig. 6.8 Clinical applications of the robot teleoperation in FAHZU for COVID-19, reprinted from Ref. [7], CC BY 4.0. **a** Medicine delivery. **b** Remote auscultation. **c** Remote operation of the medical instruments. **d** Remote daily consultation

are picked up from the storage boxes on both sides of the robot. etc.) to be delivered to the patient, thus achieving remote item delivery. The field application is shown in Fig. 6.8a;

(2) Remote operation and remote auscultation: By installing Doppler ultrasound auscultation equipment at the end of the robot's robotic arm, medical staff can remotely control the auscultation equipment at the end of the robot to contact the auscultation part of the patient's body. The auscultation signal is transmitted to remote medical staff in real time via Bluetooth to achieve remote auscultation. The field application is shown in Fig. 6.8b;

(3) Remote operation of conventional medical instruments: By installing a capacitive stylus on the panel of conventional medical instruments that can be touched at the end of the robotic arm, the operator remotely controls the robotic arm to operate the instruments outside the isolation ward. The robotic arm can operate the physical buttons of the instrument by pressing with one finger, and operate the knob of the instrument by holding and rotating with two fingers. The field application is shown in Fig. 6.8c;

(4) Non-contact consultation and emotion monitoring: By deploying an audio and video communication system on a smart tablet integrated in front of the robot, non-contact consultation between patients and doctors is achieved. In addition, an emotion recognition system based on deep neural networks is deployed to

realize daily emotion recognition and recording analysis of patients. The field application is shown in Fig. 6.8d.

References

1. Lyu H, Yang G, Zhou H et al (2020) Teleoperation of collaborative robot for remote dementia care in home environments. IEEE J Transl Eng Health Med 8:1400510
2. Department of Economic and Social Affairs PD (2019) World population prospects. United Nations, New York
3. Bardaro G, Antonini A, Motta E (2022) Robots for elderly care in the home: a landscape analysis and co-design toolkit. Int J Soc Robot 14(3):657–681
4. Yang G, Lv H, Chen F et al (2018) A novel gesture recognition system for intelligent interaction with a nursing-care assistant robot. Appl Sci-Basel 8(12):2349
5. Gao A, Murphy RR, Chen W et al (2021) Progress in robotics for combating infectious diseases. Sci Robot 6(52)
6. Meeker C, Rasmussen T, Ciocarlie M (2018) Intuitive hand teleoperation by novice operators using a continuous teleoperation subspace. In: Proceedings of the IEEE international conference on robotics and automation, Brisbane, Australia, pp 5821–5827
7. Yang Y, Peng F, Wang R et al (2020) The deadly coronaviruses: the 2003 SARS pandemic and the 2020 novel coronavirus epidemic in China. J Autoimmun 109:102434
8. Gbouna ZV, Pang G, Yang G et al (2021) User-interactive robot skin with large-area scalability for safer and natural human-robot collaboration in future telehealthcare. IEEE J Biomed Health Inform 25(12):4276–4288

Chapter 7
Conclusion and Future Perspectives

Abstract This book explores the advancements in teleoperation technology, specifically focusing on human-robot pose mapping for applications in remote homecare and medical assistance. The related study addresses challenges such as improving trajectory tracking accuracy, coordinating dual arms and redundant degrees of freedom, and understanding the impact of network performance on teleoperation systems. Four key areas of research are highlighted: single-arm and dual-arm pose mapping, full-body coordinated teleoperation, and communication-reliable teleoperation. Innovative solutions, including real-time incremental pose mapping, gesture-based control, and network coordination frameworks, enhance the performance and intuitiveness of teleoperated robots. The study also evaluates the impact of network delays and proposes methods for long-distance teleoperation, such as ultra-long-distance China-Sweden connections. Future research directions involve integrating force feedback for high-precision tasks, further exploring the effects of network delays on performance, and expanding the applications to industries like nuclear, chemical, and emergency response.

Keywords Work summary · Dual-arm control · Remote homecare · Wireless control · Human-robot interaction

This part of the book comprises one chapter, which summarizes the main work of this book and presents some important future directions.

7.1 Research Conclusions

Teleoperation technology is a key practical technology in the field of robotics, with its application scenarios expanding from hazardous operational environments in the nuclear and chemical industries to high-dynamic, strong environmental interaction scenarios such as remote homecare and medical assistance tasks. Current challenges

H. Lyu et al., *Human Motion Awareness and Robot Teleoperation*, Advanced Topics in Science and Technology in China 4, https://doi.org/10.1007/978-981-96-6545-7_7

include poor end trajectory tracking accuracy in master-slave heterogeneous human-robot pose mapping, difficulty in coordinating the dual arms, chassis, and torso of redundant degree-of-freedom robots, and unclear impact of network performance on teleoperation control systems. To address these issues, this book conducts research in four areas: "single-arm pose mapping", "dual-arm pose mapping", "full-body coordinated teleoperation", and "communication-reliable teleoperation". It captures the body movements of the master operator through inertial motion capture devices and establishes pose mapping teleoperation strategies for the robot's dual arms, chassis, and humanoid torso based on the master operator's body posture data. Additionally, it designs a control network coordination framework for the pose mapping teleoperation system to enhance the system's convenience and intuitiveness, enabling care assistant robots to become the remote "second body" of medical personnel. This has practical application needs and significant promotion value. The main work and conclusions of this book are as follows:

(1) Based on wearable inertial motion capture devices to collect the operator's limb movement data, a distributed communication architecture between the dual-arm robot control interface and the inertial motion capture device was implemented using ROS, completing the direct point-to-point pose mapping between the human and the robots. To address the issues of non-uniform speed and jerky, unsmooth tracking paths in the robot's end movement process, a human-robot motion mapping method based on local path resampling constraints was proposed, achieving continuous end effector trajectory tracking mapping between the operator and the robot. Experiments were designed to evaluate the end effector trajectory tracking performance, calculating the DTW distance between the corresponding spatial curves to quantitatively assess the similarity between the trajectories. The best tracking performance was achieved when $F_{re} = 7.5$ Hz, for both simple and complex trajectories, with end trajectories tracking errors reduced to $P_{\mathrm{RMS}} = 1.05$ mm and $P_{\mathrm{RMS}} = 5.10$ mm for simple spatial linear motion trajectories and complex spatial character motion trajectories, respectively.
(2) An incremental pose mapping strategy was proposed, which calculates the relative pose transformation between the robot and the operator in real-time for the kinematic resolution of the target pose of the teleoperated robot. This approach eliminates the cumbersome initial calibration process typically required in traditional pose mapping teleoperation. Additionally, a GuLiM dual-arm pose mapping teleoperation method based on the coupling of hand gestures and limb movements was proposed. This method addresses the issue of limited operator movement range in master-slave heterogeneous human-robot pose mapping operations, enhancing the convenience and intuitiveness of teleoperation. Comparative studies were conducted to evaluate the performance differences between the GuLiM pose mapping method and the traditional DMM direct mapping method in task completion. The GuLiM mapping method showed a 46.77% improvement in placement accuracy over the DMM direct mapping method. In the pose placement accuracy assessment experiments, the average

score of the GuLiM hybrid mapping method was 69.27% higher than that of the DMM direct mapping method.

(3) To achieve teleoperation control of the redundant degree-of-freedom body motion mechanisms of an assistive care robot, intuitive teleoperation control research based on operator limb motion capture was conducted for the robot's chassis movement control and humanoid torso pose mapping control. A wearable wrist camera was designed, and a hand gesture trajectory recognition method based on background velocity feature matching was proposed, achieving classification and recognition of various gesture trajectories. With 1000 gesture samples, the average recognition accuracy under LOSO cross-validation was 97.34%. Corresponding gesture mapping strategies were designed for the robot's movement control. Furthermore, a torso pose mapping strategy based on the recognition of operator torso and lower limb movement features was developed, enabling pose mapping control of the humanoid robot's multi-degree-of-freedom torso. Finally, application research was conducted on the proposed pose mapping teleoperation technology in two scenarios: remote homecare and remote medical assistance.

(4) A semi-physical simulation test system was designed and implemented to evaluate the performance of local wireless control. Performance tests were conducted for Wi-Fi 6 and 5G network connections at different AP distances. The test results showed that the probability distribution of Wi-Fi 6 network delays above 10 ms was higher than that of the 5G network, but the average delay of the Wi-Fi 6 network was lower than that of the 5G network. In terms of control performance, the latency between the master and slave ends was significantly higher under wireless local area network connections compared to wired connections, with larger peak joint errors. The peak joint error between the master and slave ends was greater under Wi-Fi 6 wireless network connections than under 5G network connections. For wide-area wired network teleoperation, a dedicated ultra-long-distance teleoperation network was designed and built between Sweden and China, spanning 7800 km, with an average end-to-end delay of 1.230 s and an average peak error of $-2.777°$. Furthermore, a master-end speed feedforward control method based on slave-end position feedback was proposed, reducing the average peak error between the master and slave ends to $-2.199°$, a reduction of 20.80%. This achieved practical validation and application of the intercontinental ultra-long-distance teleoperation control system.

7.2 Technical Discussion

(1) A teleoperation method for mapping the upper limb movements of a human to a robotic arm was proposed, improving the trajectory tracking accuracy of the slave robotic arm and enhancing the intuitiveness and convenience of the master operator's control process.

A local path resampling pose mapping strategy was introduced, which resamples the hand trajectory of the operator periodically to plan the desired path for the robot in real-time, constraining the trajectory tracking output of the slave robot's end-effector, thus improving the smoothness and similarity of master-slave end-effector trajectory tracking during low-frequency pose data mapping output. An incremental pose mapping method was proposed, achieving equivalent pose mapping in Cartesian space by calculating the relative pose between the human and the robot's end-effector in real-time. This simplifies the initialization calibration process of pose mapping and addresses the limitation of the operator's action range in master-slave heterogeneous human-robot pose mapping. Furthermore, dual-arm human-robot pose mapping teleoperation was achieved by coupling the operator's hand gestures with limb movements.

(2) A pose mapping strategy for non-robotic-arm-type body movement mechanisms of humanoid robots was established, achieving intuitive pose mapping teleoperation control of the robot's mobile chassis and multi-degree-of-freedom torso.

A wearable wrist camera capable of extracting the operator's hand movement background features was designed, and a dynamic hand gesture trajectory recognition method for the operator was proposed. This method enables the control of the wheeled chassis of the care-assistive robot through various hand gesture commands. Additionally, a pose mapping teleoperation control method for the redundant degree-of-freedom torso of a humanoid robot was designed based on the operator's torso and lower limb movements. This method overcomes the traditional pose mapping teleoperation output control range, achieving full-body coordinated pose mapping teleoperation for non-robotic-arm-type body movement mechanisms of care-assistive robots. Application functionality verification was conducted using a self-built care-assistive robot platform for two scenarios: remote homecare and remote medical assistance.

(3) A performance evaluation framework for pose mapping teleoperation systems was established, revealing the impact of wireless local area networks and wide-area wired networks on control performance, and achieving intercontinental ultra-long-distance pose mapping teleoperation.

A semi-physical simulation test system for evaluating the performance of pose mapping teleoperation systems was designed, and real network environments were introduced for deployment to perform real-time performance testing of teleoperation systems under various wireless local area network connections, including Wi-Fi 6 and 5G. This revealed the impact of wireless local area networks on teleoperation control performance. An intercontinental wide-area network connection framework for human-robot pose mapping teleoperation was established, testing the loopback delay and real-time pose mapping control performance of intercontinental wide-area network connections. A feedforward control method for large delays in ultra-long-distance wide-area network connections was proposed, reducing the mapping error between the master and slave ends, and achieving China-Sweden intercontinental wide-area network ultra-long-distance human-robot pose mapping teleoperation.

7.3 Look into the Future

This book focuses on the intuitive operation and convenient interaction requirements during teleoperation, investigating the human-robot pose mapping mechanism and unilateral teleoperation intelligent interaction control methods. Preliminary research was also conducted on the relationship and collaborative design between the performance of the pose mapping teleoperation system and network connection conditions. Based on the problems encountered during the research and writing process, the following prospects and discussions for future research are proposed:

(1) In practical applications, unilateral teleoperation control lacks feedback on contact forces from the slave side. For production tasks that require high precision in end-effector force/control, relying solely on visual feedback may not meet the feedback needs of the master operator and the force control accuracy requirements of the slave side. Based on the pose mapping mechanism and control network collaborative design strategy studied in this book, future research could further enhance the functionality of the master input devices and explore the pose mapping dual-arm teleoperation control with force feedback.
(2) Due to the extreme complexity of the intercontinental communication link connections and relay nodes, this study only examined the relationship between control loop round-trip delay and control performance, without breaking down the impact of different segments of delay on control performance during intercontinental ultra-long-distance teleoperation. Future work could involve setting up dedicated intercontinental network links and, using the analysis methods proposed in this book, further study the impact of various types of delays on teleoperation control performance in intercontinental ultra-long-distance network links.
(3) The specific application research conducted in this book mainly targets weak interaction task scenarios such as homecare and medical assistance. Future research could extend the applications to other industrial fields requiring professional skill operations and remote work demands, such as nuclear industry, chemical industry hazardous waste disposal, emergency public safety incident remote handling, and remote exploration missions.

GPSR Compliance

The European Union's (EU) General Product Safety Regulation (GPSR) is a set of rules that requires consumer products to be safe and our obligations to ensure this.

If you have any concerns about our products, you can contact us on ProductSafety@springernature.com

In case Publisher is established outside the EU, the EU authorized representative is:

Springer Nature Customer Service Center GmbH
Europaplatz 3
69115 Heidelberg, Germany

Batch number: 10704307

Printed by Printforce, the Netherlands